2nd International Conference on Futuristic Trends in Materials and Manufacturing 2019 (ICFTMM 2019)

IOP Conference Series: Materials Science and Engineering
Volume 802

Greater Noida, India
8 - 9 November 2019

Editors:

Kuldeep Panwar
Inderdeep Singh
Pramendra Kumar Bajpai
Satyendra Singh

ISBN: 978-1-7138-2673-6
ISSN: 1757-8981

Printed from e-media with permission by:

Curran Associates, Inc.
57 Morehouse Lane
Red Hook, NY 12571

Some format issues inherent in the e-media version may also appear in this print version.

Institute of Physics
Dirac House, Temple Back
Bristol BS1 6BE UK

Phone: 44 1 17 929 7481
Fax: 44 1 17 920 0979

techtracking@iop.org

Additional copies of this publication are available from:

Curran Associates, Inc.
57 Morehouse Lane
Red Hook, NY 12571 USA
Phone: 845-758-0400
Fax: 845-758-2633
Email: curran@proceedings.com
Web: www.proceedings.com

TABLE OF CONTENTS

Proceeding of 2nd International Conference on Futuristic Trends in Material &Manufacturing 2019

Greater Noida

(ICFTMM 2019)

8-9 November 2019

Editors:

Dr. Kuldeep Panwar, Associate Professor, DTC, GGSIPU, New Delhi

Dr. Inderdeep Singh, Professor, IIT Roorkee

Dr. Pramendra Kumar Bajpai, Associate Professor, NSUT, New Delhi

Dr. Satyendra Singh, Associate Professor, BTKIT, Dwarahat

IOP Conf. Series: Materials Science and Engineering **802** (2020) 011001 doi:10.1088/1757-899X/802/1/011001

Preface

The content of the edited book is based on papers presented at the 2nd International Conference on Futuristic Trends in Materials and Manufacturing (ICFTMM-2019) held in Delhi Technical Campus, 28/1, Knowledge Park-III, Greater Noida, Uttar Pradesh 201306 India on 8-9 November 2019. As a result of economic globalization, manufacturing enterprises face intense competition for market, resource, technology and skilled personnel. The design and manufacturing of product are becoming more and more complex. In this environment, manufacturing enterprises are seeking various integrated, just-in-time, manufacturing resources to meet the requirement of businesses and develop extensive collaboration among enterprises to cope with the rapidly changing market requirements. There are different kinds of manufacturing resources that belong to various enterprises. These enterprises are geographically distributed, morphologically diverse and autonomous, which make the resource sharing and management more complicated. This book covers in depth of manufacturing and various characterizations of different advanced materials. The domain of this book is suitable for the under graduate and post graduate students, academicians, scientists and researchers working in this field.

IOP Conf. Series: Materials Science and Engineering **802** (2020) 011002 doi:10.1088/1757-899X/802/1/011002

Peer review statement

All papers published in this volume of *IOP Conference Series: Materials Science and Engineering* have been peer reviewed through processes administered by the proceedings Editors. Reviews were conducted by expert referees to the professional and scientific standards expected of a proceedings journal published by IOP Publishing.

IOP Conf. Series: Materials Science and Engineering **802** (2020) 012001 doi:10.1088/1757-899X/802/1/012001

Modeling of Ice Slurry Flow through Bend Pipe

K.S. Rawat[1*], A. Verma[2], A. Uniyal[3], H.P. Gangwar[4], A.K. Pratihar[5]

[1,2,3]Department of Mechanical Engineering, MIET, Meerut, UP, India-250005
[4,5]Department of Mechanical Engineering, GBPUAT, Pantnagar, UK, India- 263145

kamalrawat273@gmail.com

Abstract. In this work, CFD Eulerian multiphase model with KTGFis used to investigate the flow behaviour of the ice slurry in 90° elbow bend pipe. The present model considered the particle-particle collisions along with particle-wall collisions. In the modeling of ice slurry flow various interfacial forces viz. drag force, lift force and turbulence dispersion force are considered. The solid phase viscosity is calculated as a sum of kinetic and collision viscosity component and the turbulence in the ice slurry flow is modeled by the per phase k- model. Simulation has been performed at velocity of 1 m/s, concentration of 10% and particle size of 0.25 mm for the bend radius ratios (R/r) of 2, 2.98 and 5.6. Apart from that, effect of particle size, velocity and concentration on solid particle distribution has also been investigated. Results show that, secondary flow is generated similar to single phase flow which influences the velocity, particle distribution and pressure drop.

1. Introduction

Humans are always dependent on ice for cooling from the ancient times. In earlier time before the invention of the artificial refrigeration system natural ice was used for the food preservation and cooling. The ice from the colder areas was transported to hotter areas and mixture of ice with salt also used to attain temperatures lower than 0 °C. However, with the subsequent development of artificial refrigeration lead us to harvest ice in the form of cubes, block and flakes for cooling and other applications. The Ice has good value of latent heat during melting, however, it has slow cooling rate due to poor heat transfer characteristics. Therefore, the use of ice has become superseded due low heat transfer characteristics and complications in transportation [1]. Apart from that, melting of ice creates some sharp edges due to which manual handling and transportation of ice is not too easy. Moreover, these sharp edges also responsible for damage of surfaces of product in case of direct contact chilling. The above-mentioned difficulties can be simply controlled by using ice in the form of slurry which has excellent transportation and heat transfer characteristics [2].

Ice slurry can be defined as a mixture of micro sized ice particles with binary solution of water. On the bases of use, configuration of slurry may vary i.e. the particles concentration (10-30 %), size (0.1-1.0 mm) and concentration of depressant in water. Ice slurry has good fluidity and high energy storage capacity. Apart from that it remains at constant temperature during the heat transfer due to phase change process, hence higher heat transfer coefficient in comparison to other conventional fluid. The flow of ice slurry through pipe is a convolute phenomenon in compare to other conventional fluid which affected by configuration of slurry, flow parameters and pipe geometry. For such types of problem, CFD analysis can delivers plenty information about the flow [3].

Most of the studies on ice slurry flow available in the literature are related to the determination of rheological behaviour and pressure drop. The ice slurry behaves as a Newtonian fluid for lower than 10 % initial ice concentration and non-Newtonian fluid for higher initial ice concentration [4]. Several researchers considered the different rheological model (Bingham, Power Law, Herschel–Bulkley and Cassion) for ice slurry. Monteiro and Bansal [5] proposed the empirical or semi-empirical correlations based on the mentioned rheological models to predict the pressure drop. However, these models usually considered uniform distribution of ice particles which is possible only at very high velocity therefore, fail to predict the pressure drop [6]. Vuarnoz et al. [7] and Stamatiou and Kawaji [8] measured the velocity and ice concentration distributions for ice slurry flow in horizontal and vertical pipes, respectively. However, only few experimental data are available on the velocity and concentration distribution due to complexity of the measurements.
CFD approach has been widely used by several researchers to simulate the different material's slurry flow [9-16].
In any pipe line system, bends are compulsory as they provide flexibility. During the flow through the bend pipe, secondary flow occurs as a result of centrifugal force acting on the fluid and cause additional pressure losses [17]. The generated secondary flow does not only affect the pressure drops but, also the solid particles and velocity distribution across the cross section. During the flow through the bend, the flow depends on bend radius and other operating parameters. Therefore, it is essential to investigate the flow of ice slurry through bend to develop better understanding of flow behaviour.

2. Problem Description

In this work, flow has been presented in bend pipe (as shown in Fig. 1) of 23 mm diameter with different bend radius ratios (R/r) of 2, 2.98 and 5.6, where, r is pipe radius and R is the bend radius. Pipe has contains 90° elbow bend section, downstream and upstream pipe. The investigations have been carried out in bend section.

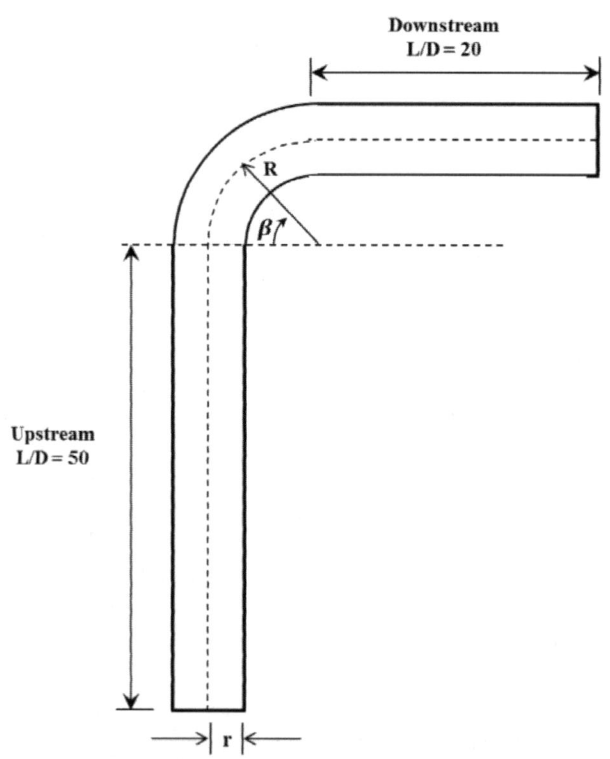

Fig. 1. Schematic diagram of 90° elbow bend pipe

3. Mathematical Modelling

The mathematical model has been developed on the bases of Eulerian model and per phase k-e turbulence model [3].

Continuity equations: Equation (1) is for liquid, (2) for solid phase where subscripts l & s denotes the liquid & solid respectively.

$$\frac{\partial}{\partial t}(\alpha_l \rho_l) + \nabla \cdot (\alpha_l \rho_l \vec{v}_l) = 0 \qquad (1)$$

$$\frac{\partial}{\partial t}(\alpha_s \rho_s) + \nabla \cdot (\alpha_s \rho_s \vec{v}_s) = 0 \qquad (2)$$

$$\alpha_l + \alpha_s = 1 \qquad (3)$$

where ρ, α, and v are density, volume fraction and velocity respectively.

Momentum Equations: Equations (4 & 5) is for liquid, (6 & 7) for solid phase

$$\frac{\partial}{\partial t}(\alpha_l \rho_l \vec{v}_l) + \nabla \cdot (\alpha_l \rho_l \vec{v}_l \vec{v}_l) = -\alpha_l \nabla P + \nabla \tau_l + \alpha_l \rho_l \vec{g} + \vec{R}_{sl} \qquad (4)$$

$$\tau_l = \alpha_l \mu_l \left(\nabla \vec{v}_l + \nabla \vec{v}_l^{T}\right) + \alpha_l \left(\lambda_l - \frac{2}{3}\mu_l\right)(\nabla \vec{v}_l)\bar{\bar{I}} \qquad (5)$$

$$\frac{\partial}{\partial t}(\alpha_s \rho_s \vec{v}_s) + \nabla \cdot (\alpha_s \rho_s \vec{v}_s \vec{v}_s) = -\alpha_s \nabla P + \nabla \tau_s + \alpha_s \rho_s \vec{g} - \nabla P_s + \vec{R}_{ls} \qquad (6)$$

$$\tau_s = \alpha_s \mu_s \left(\nabla \vec{v}_s + \nabla \vec{v}_s^{T}\right) + \alpha_s \left(\lambda_s - \frac{2}{3}\mu_s\right)(\nabla \vec{v}_s)\bar{\bar{I}} \qquad (7)$$

where μ, R, $\bar{\bar{I}}$ and λ_s are viscosity, interfacial forces, unit tensor and bulk viscosity respectively. Bulk viscosity defined from Lun et al., 1984 **[12]**.

$$\lambda_s = \frac{4}{3}\alpha_s \rho_s d_s g_{o,ss}(1 + e_{ss})\left(\frac{\Theta}{\pi}\right)^{0.5} \qquad (8)$$

The interfacial forces $(\vec{R})$ includes lift and drag forces **[13]**. Drag force is described by Gidaspow, 1994 **[14]**. The lift force on the solid phase is calculate in the present work as

$$\vec{F}_{lift,sl} = C_L \rho_l \alpha_s (\vec{v}_l - \vec{v}_s) \times \nabla \vec{v}_l \qquad (9)$$

$$\vec{F}_{lift,sl} = -\vec{F}_{lift,ls} \qquad (10)$$

Lift force coefficient (C_L) is assigned to be 0.2.

Solid viscosity is the sum of kinetic and collisional viscosity. Component of frictional viscosity is zero.

$$\mu_s = \mu_{col} + \mu_{kin} + \mu_{fri} \qquad (11)$$

$$\mu_{col} = \frac{4}{5}\alpha_s \rho_s d_s g_{o,ss}(1 + e_{ss})\left(\frac{\Theta}{\pi}\right)^{0.5} \qquad (12)$$

$$\mu_{kin} = \frac{\alpha_s \rho_s d_s \sqrt{\pi \, \Theta_s}}{6(3 - e_{ss})}\left[1 + \frac{2}{5}(1 + e_{ss})(3e_{ss} - 1)\alpha_s g_{o,ss}\right] \qquad (13)$$

$$g_{o,ss} = \left[1 - \left(\frac{\alpha_s}{\alpha_{s,\max}}\right)^{1/3}\right]^{-1} \tag{14}$$

where $\alpha_{s,\max}$ is maximum packing limit.

Table 1: Boundary conditions

Boundary	Boundary Condition	Description
Inlet	Velocity Inlet	Liquid phase velocity $=v_l$ Solid phase velocity $=v_s = v_l$ Volume fraction $=\alpha_s$
Outlet	Pressure Outlet	For mixture: Atmospheric pressure
Wall	Wall	For liquid phase: No Slip For solid phase: Partial slip $\psi = 0.02, e_{sw} = 0.95$

Table 2: Liquid phase properties with 10% ethanol concentration

FPD	Ice Concentration (%)	Density (kg/m³)	Viscosity (m-Pa)
Ethanol	10	983.9	0.00458
Ethanol	30	981.3	0.00609

3.1 Numerical Procedure

The boundary conditions and properties of ice slurry have been given in Table 1 and 2 respectively. The geometry is divided into the hexahedral grids and grid is also optimized. In the present work three different bend radius ratios of 2, 2.98 and 5.6 for which optimum number of elements are 347384, 353160 and 366633 respectively.

3.2 Validation

Model has been validated with the experimental result of Kaushal et al. [17]. The variation in the experimental and simulation values is within 10%.

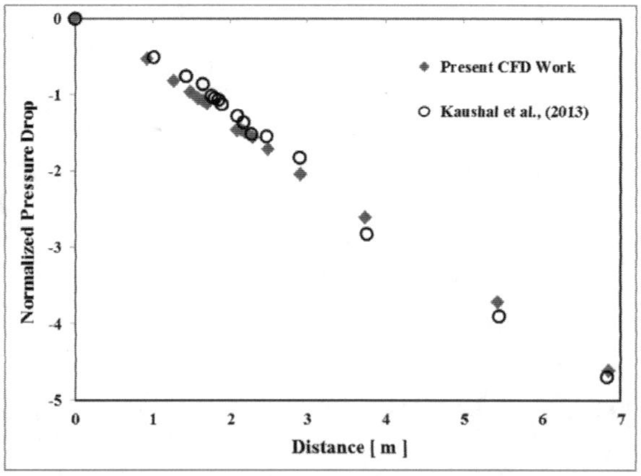

IOP Conf. Series: Materials Science and Engineering **802** (2020) 012001 doi:10.1088/1757-899X/802/1/012001

Fig. 2.Comparison between the present results and experimental data [17]

4. Results and Discussion

The simulation is performed for at the velocity of 1 m/s and concentration of 10 %. Fig. 3 shows the variation in the pressure, in plane which contained 20D length of upstream pipe from β=0°, elbow bend section and 20D length of downstream pipe from β=90°. It is shown in figure that pressure varies only in longitude direction in both upstream and downstream part and in the bend its variation is in both directions (longitudinal and radial direction) because of the centrifugal force.

It can also be observed that pressure is considerably reduced in the inner side of the bend. As a result secondary flow is generated (as shown in Fig. 4).

The boundary layer separation has also been observed at the velocity of 1 m/s and concentration of 10 % for R/r = 2. It can be observed from the fig.5 that boundary layer separation occurs and then reattached.

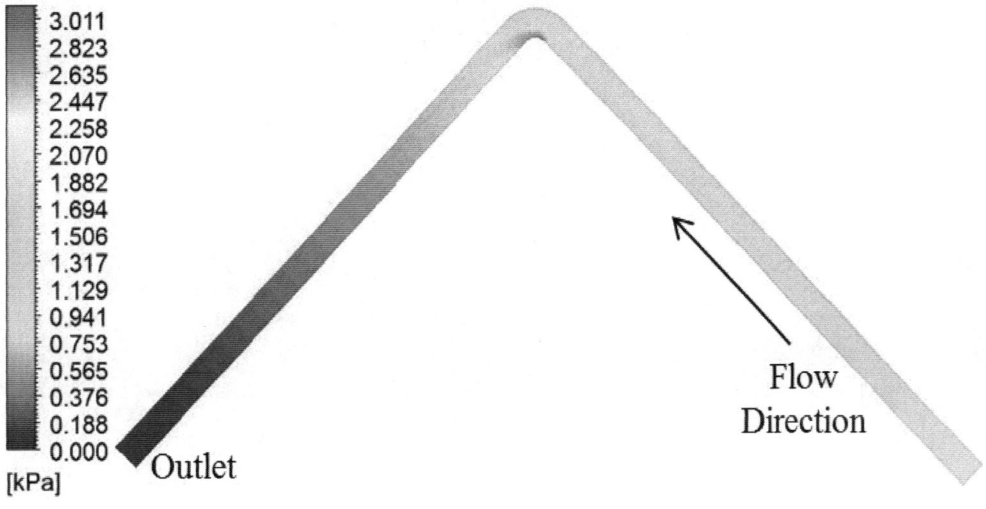

Fig. 3.Contour of pressure in a plane of bend

IOP Conf. Series: Materials Science and Engineering **802** (2020) 012001 doi:10.1088/1757-899X/802/1/012001

Fig. 4.Vector Plot of velocity in a plane of bend

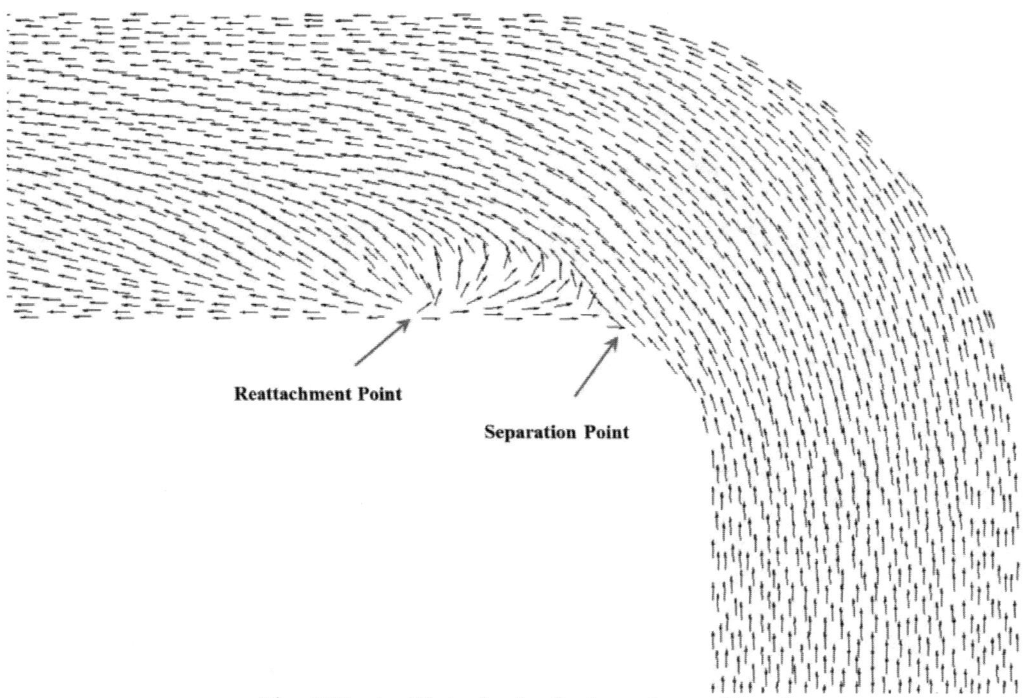

Fig. 5.Vector Plot of velocity in a plane of bend

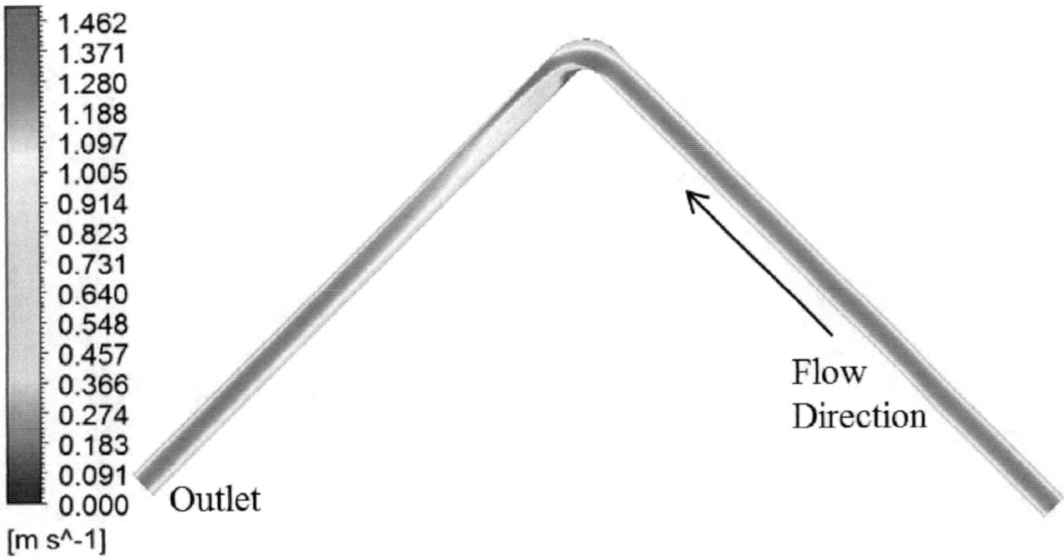

Fig. 6. Contour of velocity in a plane of bend

Fig. 6 shows the velocity contour and from the figure it can be observed that flow is fully developed earlier elbow section and it get disturbed in bend section then again reattributed. The velocity contour shows that the velocity varies dramatically in the bend. The velocity first increase near the inner side of the bend and after 45° velocity decrease and increase at the outer bend side. It is clearly shown in the figure that, flow is redistribution in the downstream section and flow becomes steady.

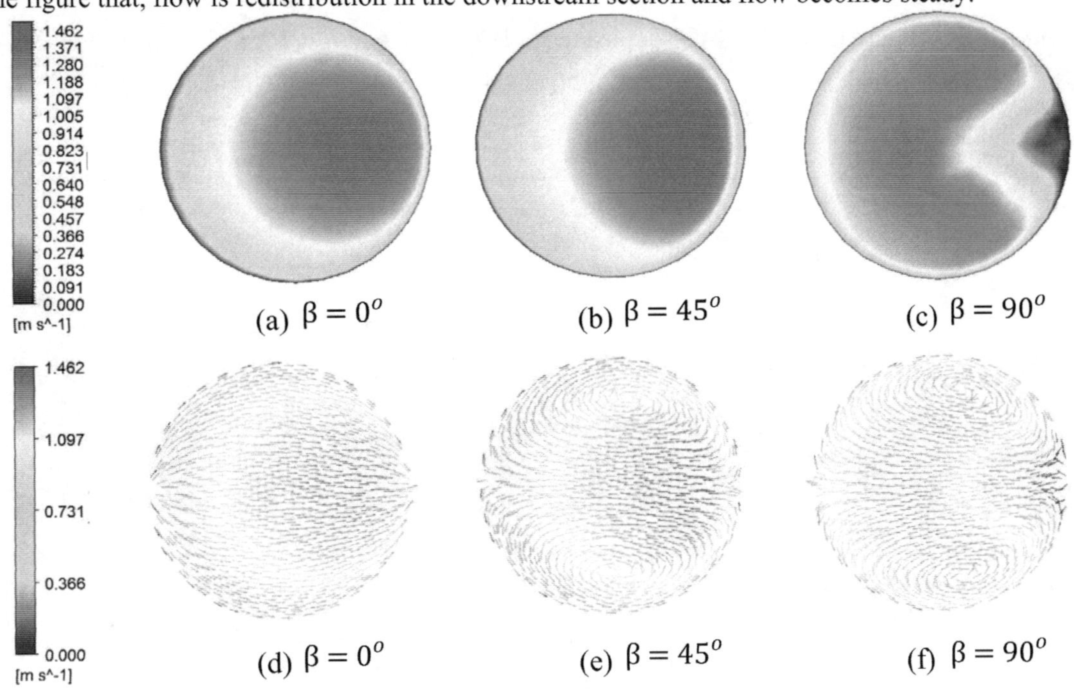

Fig. 7. Contours (a) to (c) and vector plots (d) to (f) of solid phase velocity

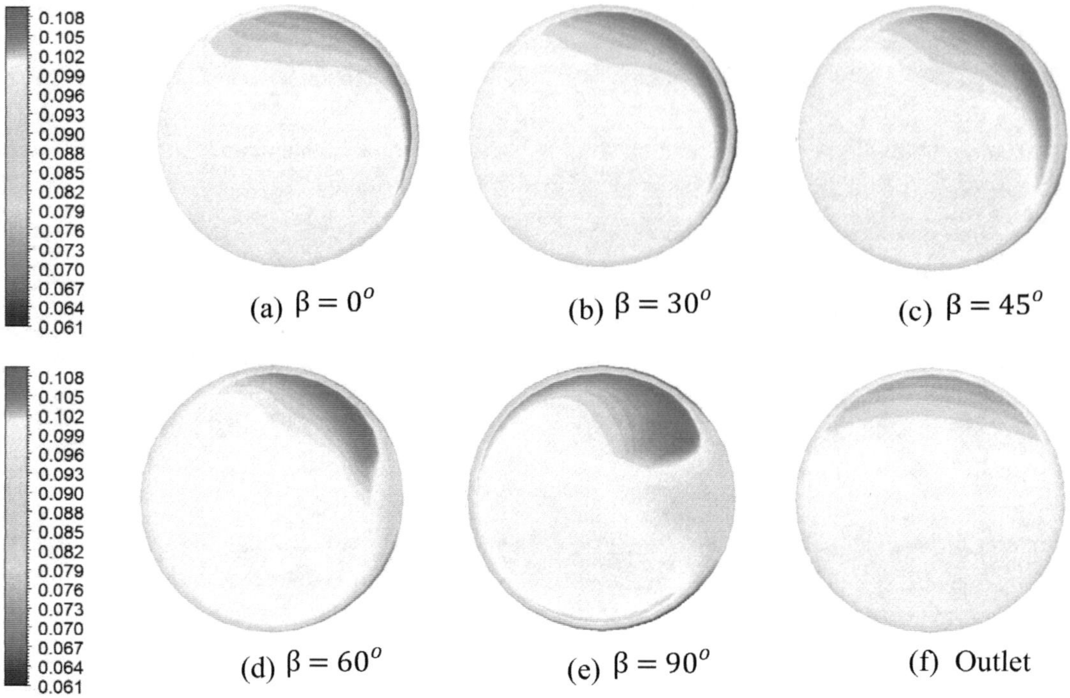

Fig. 8. Contours of particle concentration at different bend angles

The contours of velocity (a to c) and vector plots (d to f) are shown in fig.7 for V = 1 m/s, IPC = 10% and ds = 0.25 mm. The results are shown for the plane at different bend angles ($\beta = 0^o, 45^o \& 90^o$). At the inlet of the bend section the velocity increase near the bend section due to small path (shown in contour (a) and vector plot (d)). The contour (b) and vector plot (e) shows that velocity at bend angle of 45°. It can be observed that velocity is still maximum near to inner side however, particles are moving towards outer side. Due to the secondary flow velocity increases at the outer side and decreases near the inner side of the bend at the 90° bend angle.

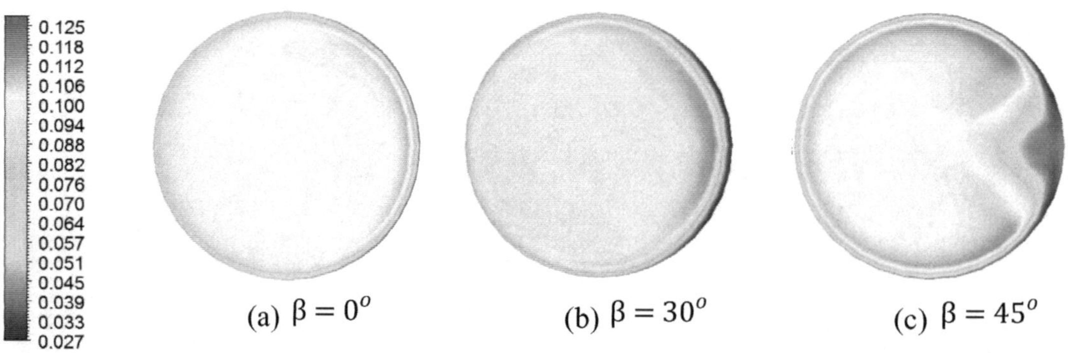

V = 3 m/s, IPC = 10%, ds = 0.25 mm

IOP Conf. Series: Materials Science and Engineering **802** (2020) 012001 doi:10.1088/1757-899X/802/1/012001

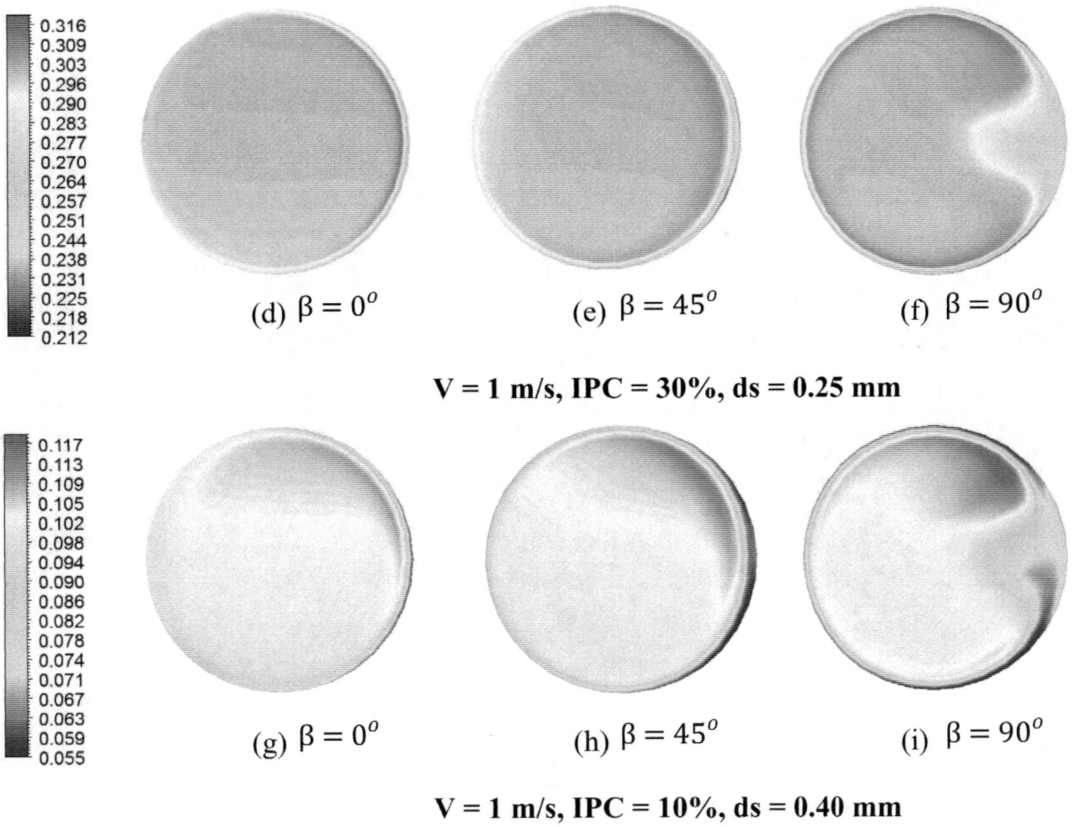

Fig. 9. Contours of particle concentration at different bend angles for different operating conditions

For the velocity of 1 m/s, concentration 10% and 10 % ethanol based slurry with 0.25 mm diameter results are shown in Fig. 8. The contours of concentration have been shown at different bend angle (a to e) and outlet for elbow pipe with bend radius ratio of 2. Figure shows that at the $\beta = 0^{o}$ flow is heterogeneous. The concentration of particles increase toward the inner side upto $\beta = 45^{0}$ and after that decrease. At the outlet of pipe, flow is more uniform due to redistribution. The effect of particle wall collision can also be observed from the figure.
Fig. 9 shows the concentration contour at $\beta = 0^{o}$, 45° and 90° for different conditions for bend radius ratio 2. The contours (a) to (c) shows the concentration contours for velocity 3 m/s, concentration 10% and particle 0.25 mm of size. Contour shows that at higher velocity flow is more uniform and effect of particle wall collision is more significant. The contours (d) to (f) shows the concentration contours for velocity 1 m/s, concentration 30% and particle 0.25 mm of size. Contour shows that at higher concentration flow is more homogeneous due to more collisions. The contours g) to (i) shows the concentration contours for velocity 1 m/s, concentration 10% and particle 0.40 mm of size. Contour shows that with bigger particles flow is more heterogeneous due to more buoyancy force.

IOP Conf. Series: Materials Science and Engineering **802** (2020) 012001 doi:10.1088/1757-899X/802/1/012001

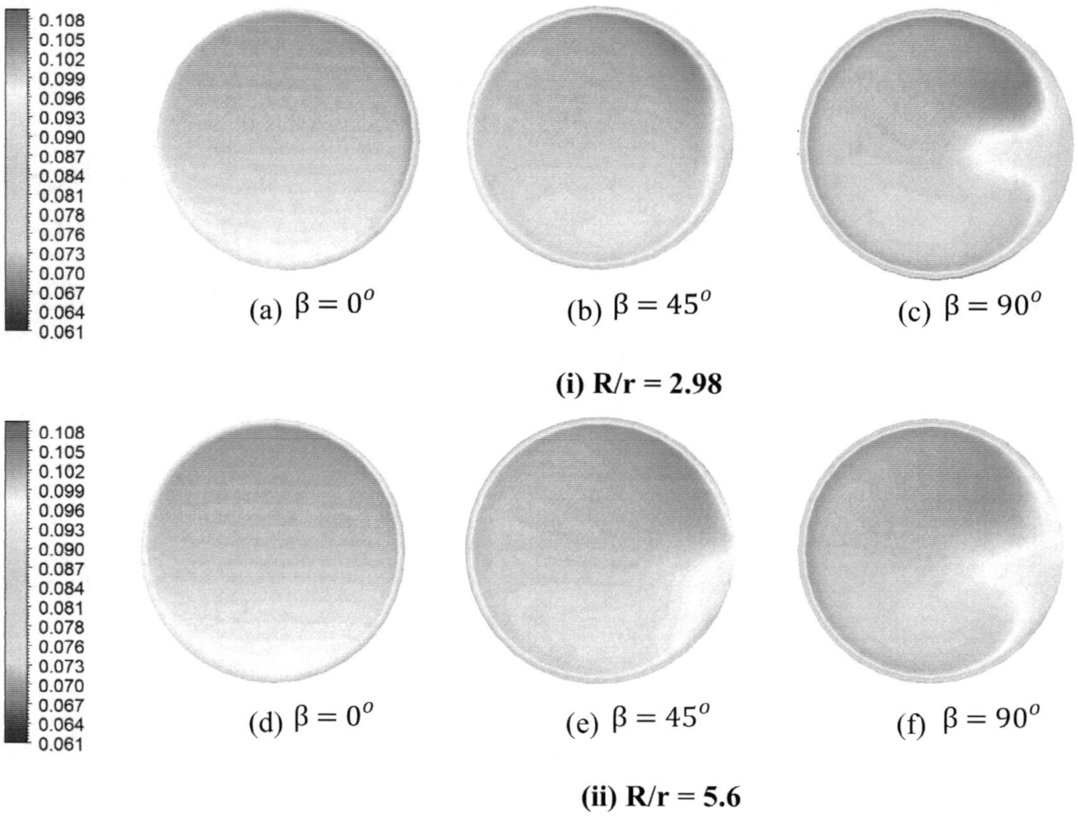

Fig. 10.Contours of particle concentration at different bend angle

Fig. 5 shows the boundary layer separation for R/r = 2. However, for the higher R/r (2.98 & 5.6) boundary layer separation does not occur. Fig. 10 shows the comparison of concentration contour at velocity 1 m/s, concentration 10% and 0.25 mm size for two different R/r. The contours (a) to (c) and contours (d) to (f) are showing the solid concentration contour at $\beta = 0^o, 45^o$ & 90° for R/r of 2.98 and 5.6, respectively. It is clear from contour that effect of secondary flow become inferior for R/r = 5.6. Apart from that, pressure losses also decrease to 4% and 14% with increase in the bend pipe radius ratio from 2 to 2.98 and 5.60.

5. Conclusions

From the simulations it is concluded that, during the flow through bend, a centrifugal force acts on the fluid which results in secondary flow at the bend section. Due to the centrifugal force acting on the fluid in the bend section, a secondary flow is generated which may also lead to boundary layer separation. For the ice slurry flow at 1 m/s and ice concentration of 10% in the 90° elbow bend pipe with bend radius (R/r) ratio of 2, the boundary layer separation occurs (shown in Fig. 5), however, for the higher (R/r) ratios of 2.98 and 5.60 boundary layer separation does not occur. The effect of secondary flow becomes inferior at higher bend radius ratios. As a result pressure losses across the pipe decrease to 4% and 14 % with increase in the bend radius ratio from 2 to 2.98 and 5.60, respectively.

The secondary flow also affects the velocity and particle distribution. The velocity of the particle first increased near the inner wall of the bend at the inlet of the bend (0° bend angle) and beyond the 45° bend angle the velocity of the particle decreases rapidly near the inner wall and increases towards the outer wall of the bend. However, the velocity distribution again becomes steady in the downstream pipe due to redistribution of particles and fluid. At the higher initial velocity, the effect of secondary flow is more significant therefore, particle distribution is more influenced by secondary flow at bend section. However, particle distribution is more homogeneous at higher velocity and higher concentration at the outlet of the pipe.

References

[1] Kauffeld M, Wang MJ, Goldstein V, Kasza KE. Ice slurry applications. International Journal of Refrigeration. 2010 Dec 31;33(8):1491-505.
[2] Rawat, K.S. and Pratihar, A.K., 2018. Thermo-Physical Properties of Aqueous Solution of Ice Slurry. International Journal of Pure and Applied Mathematics, 119(17), pp.1493-1497.
[3] Rawat, K.S. and Pratihar, A.K., 2018, February. Numerical Investigation of Ice Slurry Flow in a Horizontal Pipe. In IOP Conference Series: Materials Science and Engineering (Vol. 310, No. 1, p. 012095). IOP Publishing.
[4] Grozdek, M., Khodabandeh, R., &Lundqvist, P. (2009). Experimental investigation of ice slurry flow pressure drop in horizontal tubes. Experimental Thermal and Fluid Science, 33(2), 357-370.
[5] Monteiro, A. C., &Bansal, P. K. (2010). Pressure drop characteristics and rheological modeling of ice slurry flow in pipes. International Journal of Refrigeration, 33(8), 1523-1532.
[6] Wang, J., Wang, S., Zhang, T., &Battaglia, F. (2017). Mathematical and experimental investigation on pressure drop of heterogeneous ice slurry flow in horizontal pipes. International Journal of Heat and Mass Transfer, 108, 2381-2392.
[7] Vuarnoz, D., Sari, O., Egolf, P. W., &Liardon, H. (2002, September). Ultrasonic velocity profiler UVP-XW for ice-slurry flow characterisation. In Proceedings of the third international symposium on ultrasonic Doppler method for fluid mechanics and fluid engineering, Lausanne, Switzerland.
[8] Stamatiou, E., &Kawaji, M. (2005). Thermal and flow behavior of ice slurries in a vertical rectangular channel. Part I: Local distribution measurements in adiabatic flow. International Journal of Heat and Mass Transfer, 48(17), 3527-3543.
[9] Ekambara, K., Sanders, R. S., Nandakumar, K., &Masliyah, J. H. (2009). Hydrodynamic simulation of horizontal slurry pipeline flow using ANSYS-CFX. Industrial & Engineering Chemistry Research, 48(17), 8159-8171.
[10] Kaushal, D. R., Thinglas, T., Tomita, Y., Kuchii, S., & Tsukamoto, H. (2012). CFD modeling for pipeline flow of fine particles at high concentration. International Journal of Multiphase Flow, 43, 85-100.
[11] Kumar Gopaliya, M., &Kaushal, D. R. (2016). Modeling of sand-water slurry flow through horizontal pipe using CFD. Journal of Hydrology and Hydromechanics, 64(3), 261-272.
[12] Zhang, P., & Jiang, Y. Y. (2014). Forced convective heat transfer of slush nitrogen in a horizontal pipe. International Journal of Heat and Mass Transfer, 71, 158-171.
[13] Shi, X. J., & Zhang, P. (2016). Conjugated heat and mass transfer during flow melting of a phase change material slurry in pipes. Energy, 99, 58-68.
[14] Jin, T., Li, Y. J., Liang, Z. B., Lan, Y. Q., Lei, G., &Gao, X. (2017). Numerical prediction of flow characteristics of slush hydrogen in a horizontal pipe. International Journal of Hydrogen Energy, 42(6), 3778-3789.
[15] Nayak, B. B., &Chatterjee, D. (2017). Numerical investigation of convective heat transfer in

pipeline flow of multi-sized mono dispersed fly ash-water slurry. International Journal of Heat and Mass Transfer, 108, 1802-1818.

[16] Singh, J. P., Kumar, S., &Mohapatra, S. K. (2017). Modelling of two phase solid-liquid flow in horizontal pipe using computational fluid dynamics technique. International Journal of Hydrogen Energy, 42(31), 20133-20137.

[17] Kaushal, D. R., Kumar, A., Tomita, Y., Kuchii, S., & Tsukamoto, H. (2013). Flow of mono-dispersed particles through horizontal bend. *International Journal of Multiphase Flow*, *52*, 71-91

IOP Conf. Series: Materials Science and Engineering **802** (2020) 012002 doi:10.1088/1757-899X/802/1/012002

Numerical Investigation on Critical Velocity of Ice Slurry Flow

K.S. Rawat[1*], A. Verma[2], A.K. Pratihar[3], A.K. Singh[4], P. Maheshwari[5]

[1,2]Department of Mechanical Engineering, MIET, Meerut, UP, India-250005
[3,4,5]Department of Mechanical Engineering, GBPUAT, Pantnagar, UK, India- 263145

kamalrawat273@gmail.com

Abstract. The use of ice slurry as a working fluid in secondary-loop of refrigeration system not only enhances the system performance but also reduces the refrigerant charge and its leakage to environment. It can be transported through pipes; however, flow of ice slurry is a quite complex two phase flow, which significantly differs from conventional fluid flow. Ice slurry flow can be influenced by on several operating parameter viz. initial velocity, ice concentration, freezing point depressant concentration and particle size. Solution of such type of flow through pipe is a difficult task through experiments due to complexities in visualization and measurements. Therefore, a CFD model is developed to simulate the ice slurry flow to predict the safe conditions for the transportation of ice slurry. In the present work, numerical investigation of flow of ice slurry has been analyzed in order to predict the critical velocity and effect of different parameters on critical velocity. Eulerian granular flow model has been used for the simulation, which is also validated with existing experimental results in the literature. Result shows that, the critical velocity is especially influenced by the particle size. Critical velocity at 10 % initial ice concentration increases from 0.4 m/s to 0.16 m/s with increase in particle size from 0.1 mm to 0.4 mm.

1. Introduction

The HVAC&R systems are consuming a huge share of the entire world electricity generation and also damaging the environmental by emission of the synthetic refrigerants [1]. It is reported in the literature that, use of the SLR system enhances performance and reduce the refrigerant charge [2]. The SLR systems consist two loops: primary loop and secondary loop. The primary loop is a conventional vapour compression refrigeration system in which refrigerant flow through the expansion valve, compressor, condenser, and evaporator. The secondary loop is connected with a primary loop through the evaporator. The working fluid in the SLR system can be single phase fluid or a two phase fluid [2-3]. The single phase fluids are the low freezing point solutions which can be classified in aqueous and non-aqueous solutions. The aqueous solutions are more preferred over non-aqueous solutions as they have much lower volume flow rate for the same cooling conditions. The two phase fluids can also use in the SLR system, which has various advantages over the single phase fluids. The two phase fluid can be generated by forming of slurry of PCM. PCM slurry can be defined as the mixture of solid PCM particles with liquid and which can be flow through pipes [4-7]. Ice slurry is a PCM slurry, which is mixture of aqueous solution and ice. However, during transportation its flow is significantly differs from the conventional fluid flow [8]. Ice slurry flow influenced by different parameter i.e. inlet velocity, FPD concentration, ice concentration, pipe and particle size.

IOP Conf. Series: Materials Science and Engineering **802** (2020) 012002 doi:10.1088/1757-899X/802/1/012002

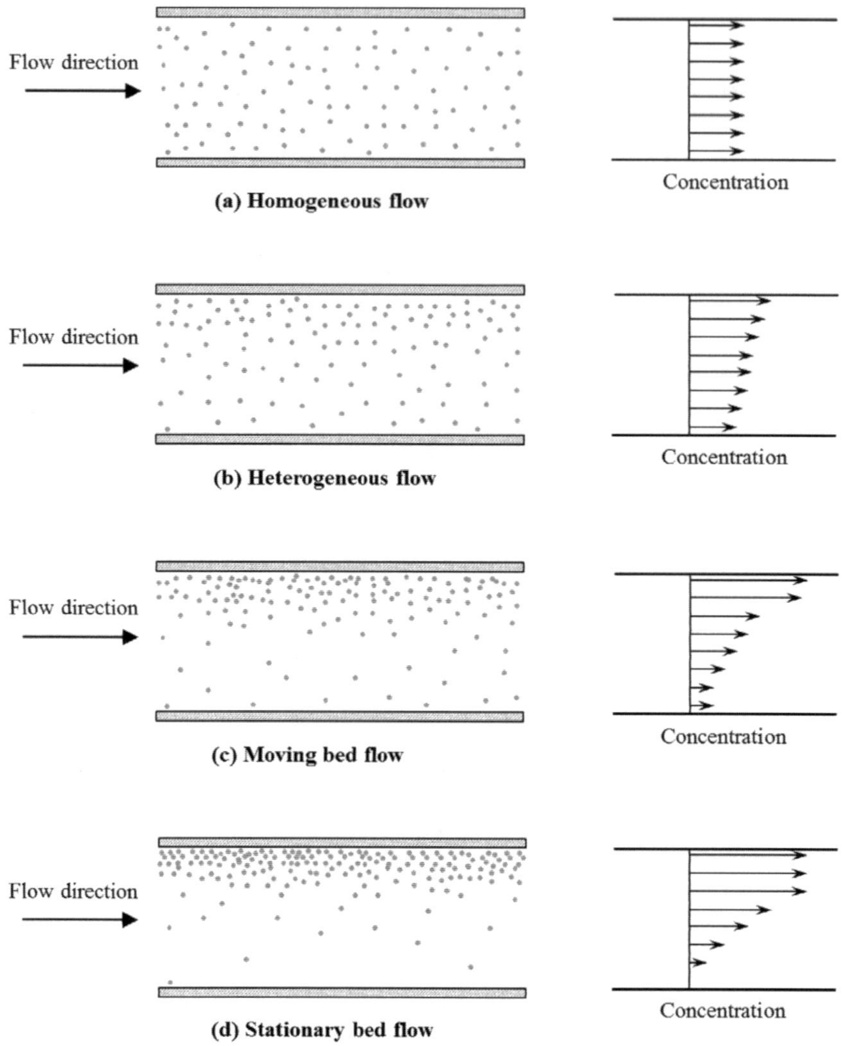

Fig. 1. Different flow regimes during the flow of ice slurry

During the flow, the ice particles tend to settle (due to density difference) towards the top of the pipe, as a result different concentration lies in the radial direction [9]. Depending on concentration distribution, flow can be classified into homogeneous, heterogeneous, moving bed and stationary bed flow regimes (shown in Fig. 1).

At a very high flow velocity, particles are uniformly suspended and flow regime is known as the homogeneous flow and if the particles are randomly suspended such that concentration gradient is formed in flow is called heterogeneous flow. At low velocity the concentration increases at the top of the pipe and layer with maximum possible concentration forms, the corresponding velocity at this stage known as critical velocity [10]. This layer moves along the top wall and in the bottom of this layer heterogeneous flow exists. In such type of concentration distribution, flow known as moving bed flow. Further reduction in velocity leads to increase in the thickness of layer and at a very low velocity, ice layer become stationary and only liquid flow through bottom side known as stationary bed condition [11].During the transportation, to avoid chock condition, the velocity should be far away from critical velocity which is challenging task by experiments. Therefore, in this section simulation of the ice slurry flow at a low initial velocity has been carried out to predict the value of critical velocity and the effect of different parameters on critical velocity.

2. Mathematical Modelling

The mathematical model has been developed on the bases of Eulerian approach.
Continuity equations: Equation (1) is for liquid, (2) for solid phase where subscripts l & s denotes the liquid & solid respectively.

$$\frac{\partial}{\partial t}(\alpha_l \rho_l)+\nabla\cdot(\alpha_l \rho_l \vec{v}_l)=0 \quad (1)$$

$$\frac{\partial}{\partial t}(\alpha_s \rho_s)+\nabla\cdot(\alpha_s \rho_s \vec{v}_s)=0 \quad (2)$$

$$\alpha_l+\alpha_s=1 \quad (3)$$

where ρ, α, and v are density, volume fraction and velocity respectively.

Momentum Equations: Equations (4 & 5) is for liquid, (6 & 7) for solid phase

$$\frac{\partial}{\partial t}(\alpha_l \rho_l \vec{v}_l)+\nabla\cdot(\alpha_l \rho_l \vec{v}_l \vec{v}_l)=-\alpha_l \nabla P+\nabla \tau_l+\alpha_l \rho_l \vec{g}+\vec{R}_{sl} \quad (4)$$

$$\tau_l=\alpha_l \mu_l\left(\nabla \vec{v}_l+\nabla \vec{v}_l^{T}\right)+\alpha_l\left(\lambda_l-\frac{2}{3}\mu_l\right)(\nabla \vec{v}_l)\overline{\overline{I}} \quad (5)$$

$$\frac{\partial}{\partial t}(\alpha_s \rho_s \vec{v}_s)+\nabla\cdot(\alpha_s \rho_s \vec{v}_s \vec{v}_s)=-\alpha_s \nabla P+\nabla \tau_s+\alpha_s \rho_s \vec{g}-\nabla P_s+\vec{R}_{ls} \quad (6)$$

$$\tau_s=\alpha_s \mu_s\left(\nabla \vec{v}_s+\nabla \vec{v}_s^{T}\right)+\alpha_s\left(\lambda_s-\frac{2}{3}\mu_s\right)(\nabla \vec{v}_s)\overline{\overline{I}} \quad (7)$$

where μ, R, $\overline{\overline{I}}$ and λ_s are viscosity, interfacial forces, unit tensor and bulk viscosity respectively. Bulk viscosity defined from Lun et al., 1984 [12].

$$\lambda_s=\frac{4}{3}\alpha_s \rho_s d_s g_{o,ss}(1+e_{ss})\left(\frac{\Theta}{\pi}\right)^{0.5} \quad (8)$$

The interfacial forces $(\vec{R})$ includes lift and drag forces **[13]**. Drag force is described by Gidaspow, 1994 **[14]**. The lift force on the solid phase is calculate in the present work as

$$\vec{F}_{lift,sl}=C_L \rho_l \alpha_s(\vec{v}_l-\vec{v}_s)\times\nabla\vec{v}_l \quad (9)$$

$$\vec{F}_{lift,sl}=-\vec{F}_{lift,ls} \quad (10)$$

Lift force coefficient (C_L) is assigned to be 0.2.
Solid viscosity is the sum of frictional, kinetic and collisional viscosity however, frictional viscosity contributes when concentration reaches the frictional limit.

$$\mu_s=\mu_{col}+\mu_{kin}+\mu_{fri} \quad (11)$$

$$\mu_{col}=\frac{4}{5}\alpha_s \rho_s d_s g_{o,ss}(1+e_{ss})\left(\frac{\Theta}{\pi}\right)^{0.5} \quad (12)$$

$$\mu_{kin}=\frac{\alpha_s \rho_s d_s \sqrt{\pi\,\Theta_s}}{6(3-e_{ss})}\left[1+\frac{2}{5}(1+e_{ss})(3e_{ss}-1)\alpha_s g_{o,ss}\right] \quad (13)$$

$$\mu_{fri} = \frac{P_s \sin\varphi}{2\sqrt{2I_{2D}}} \tag{14}$$

$$g_{o,ss} = \left[1 - \left(\frac{\alpha_s}{\alpha_{s,\max}}\right)^{1/3}\right]^{-1} \tag{15}$$

where $\alpha_{s,\max}$ is maximum packing limit.

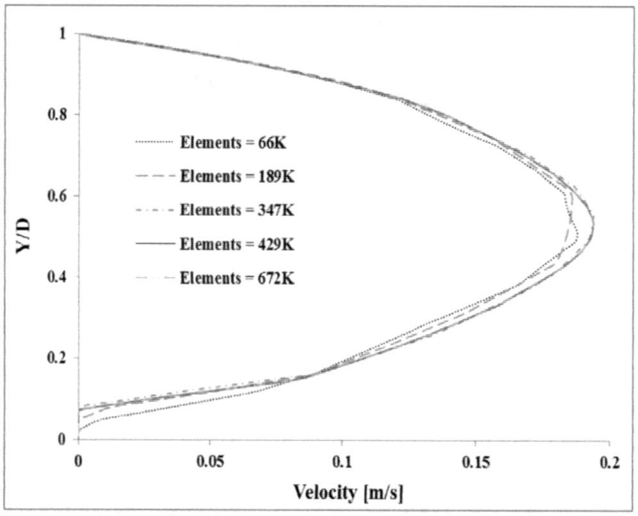

Fig. 2. Variation of solid phase velocity profile with different grid size

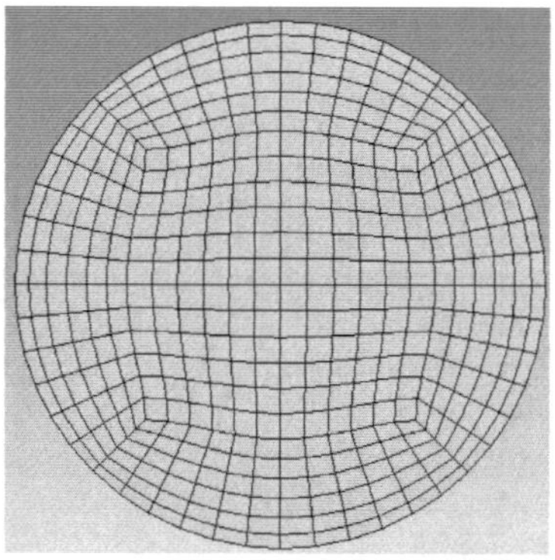

(a) Cross-section of meshed pipe

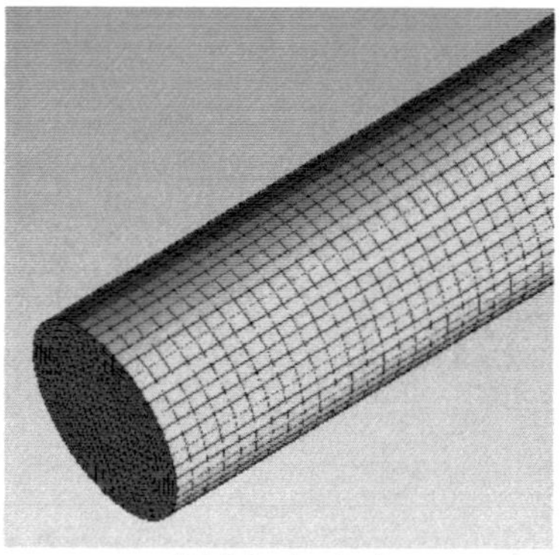

(b) Meshed pipe

Fig. 3. *Horizontal meshed pipe for laminar flow*

2.1. Discretization

Simulation has been performed in a pipe of 23 mm diameter and 2 m length. The hexahedral mesh by 'O' grid and sweep method has been generated using ICEM. At a velocity of 0.1 m/s, 10% concentration with 10 % ethanol and 0.1 mm particles, mesh independent test has been performed. The velocity profile of the solid phase has been drawn (fig. 2) that shows that, the variation for elements 347000 and 429000 is very less and it becomes negligible with further refinement. Therefore, 429000 elements are considered as optimum mesh size (shown in Fig. 3).

2.2. Boundary Conditions
The boundary conditions for each phase are assigned at inlet, outlet and wall. For the inlet, velocities and concentration are assumed to be uniform. For the outlet, pressure outlet condition, at wall, no-slip condition is adopted for liquid phases and for the solid phase Johnson–Jackson condition is adopted.

3. Validation
Present model has also been validated with the Kauffeld et al., 2005 [16]. The value of critical velocity lies between 0.13 to 0.25 m/s in an experiment. Fig.4 shows that critical velocity lies in range of 0.15 to 0.20 m/s for the same condition. Hence, present model has ability to predict the critical velocity.

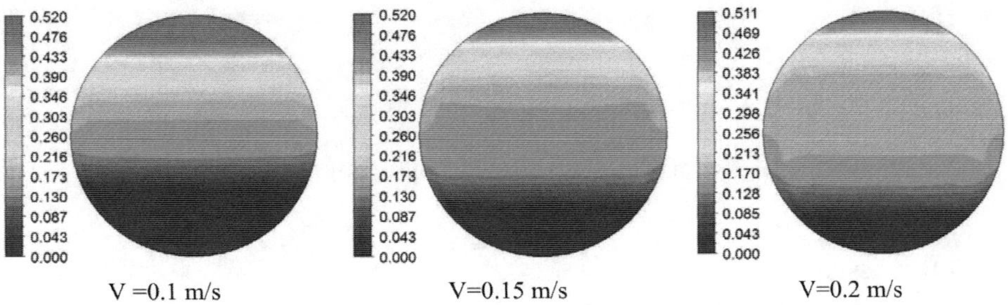

Fig. 4. Contours of the ice particles concentration at different initial velocity

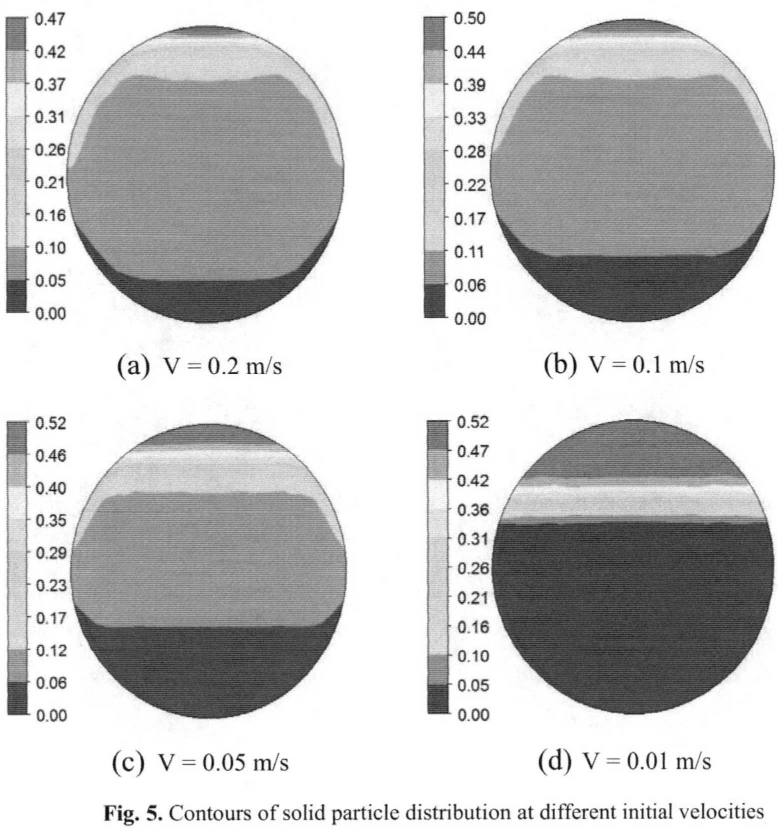

Fig. 5. Contours of solid particle distribution at different initial velocities

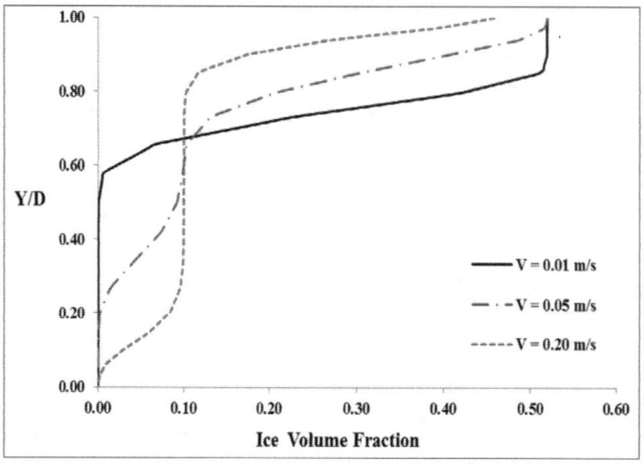

Fig. 6. Solid particle distribution profile at different initial velocities

4. Results and Discussion

Simulation has been carried out to identify the critical velocity. Fig. 5 shows the contours of the concentration for the different velocities of 0.01, 0.05, 0.01 & 0.02 m/s. The results are obtained for 10 % concentration with 0.25 mm diameter in the 10% ethanol based ice slurry [17]. It is clear from the contours (a to d) that flow converted from homogeneous to moving bed when velocity reduced from 0.1 to 0.05 m/s. The same results have also been shown in Fig. 6 except at velocity of 0.1 m/s. It can be observed from the figure that, the maximum concentration reaches to 0.52 (maximum packing limit) at velocity of 0.05 m/s.

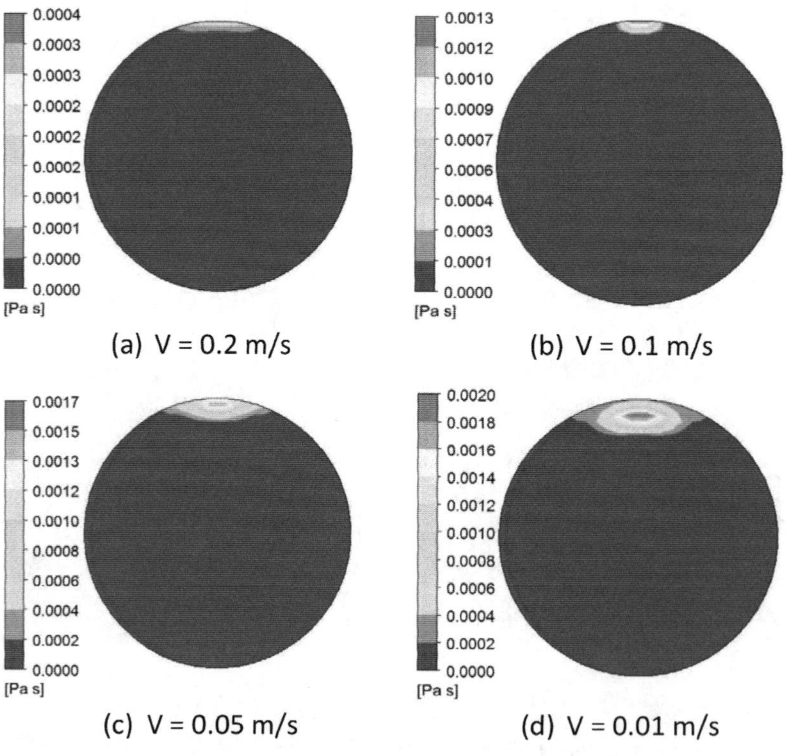

Fig. 7. Contours of frictional viscosity at different initial velocities

Fig. 7 shows the contours (a to d) of the frictional viscosity at the different velocities of 0.01, 0.05, 0.01 & 0.02 m/s. In present model the shear viscosity affected with frictional viscosity when concentration crosses the value of friction packing limit which is considered 0.45. It has been shown in the contours that the frictional viscosity reach to the significant value when concentration cross the frictional packing limit.

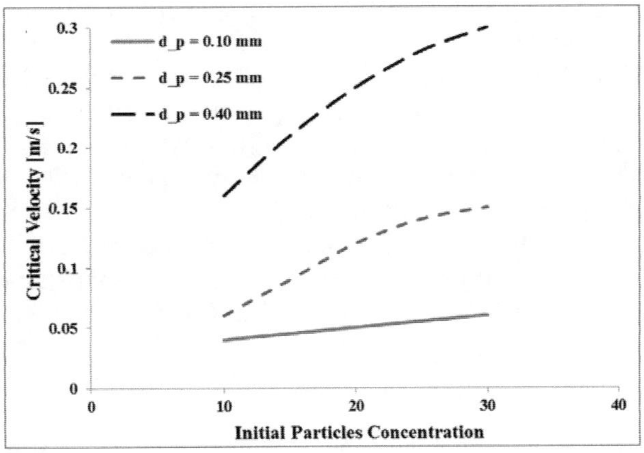

Fig. 8. Variation of the critical velocity with initial particle concentration and ice particle diameter

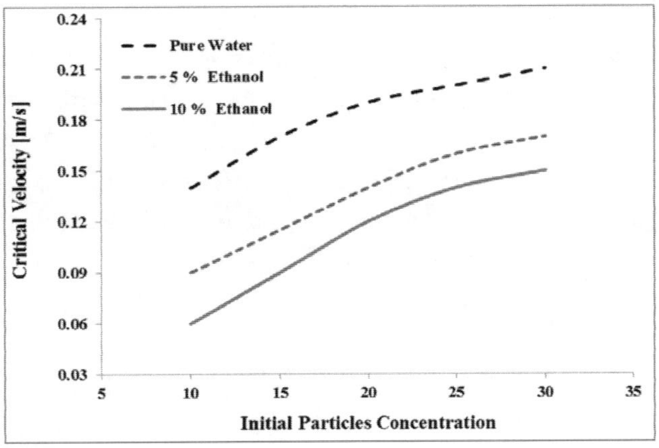

Fig. 9. Variation of the critical velocity with initial particle concentration and different concentration of ethanol

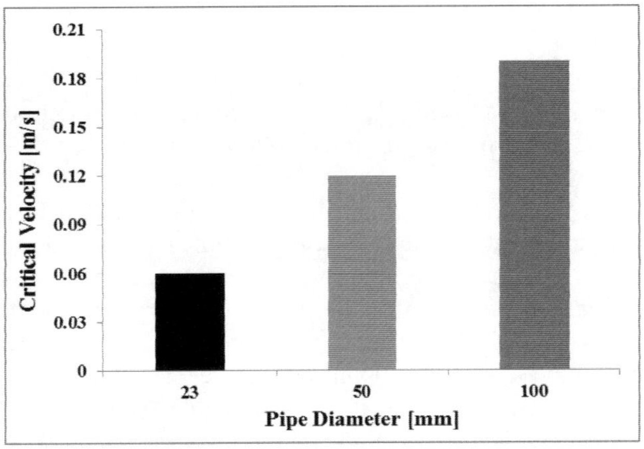

Fig. 10. Variation of the critical velocity with different pipe diameters

The effect of particle diameter and concentration on critical velocity has been shown in Fig. 8. The figure shows the concentration on the x-axis and velocity on the y-axis. It is clear from the figure that, the critical velocity increases with increase in concentration as at more concentration rate of moving bed formation increase. However, critical velocity affected most with particle diameter as the fact that bigger particle are subjected to strong buoyancy and the inertia forces in comparison to small particle. The effect of ethanol concentration on critical velocity has been shown in Fig. 9 for particle diameter of 0.25 mm. The figure shows the concentration on the x-axis and velocity on the y-axis. The results are obtained for 0%, 5% and 10% ethanol in the slurry. It can be observed from the figure that, with the decrease in ethanol concentration in the slurry, the value of critical velocity increases as density difference increase between the phases and decrease in viscosity [17].

Fig. 10 shows the value of critical velocity for different diameter of pipes. Three different pipe diameters 23, 50 and 100 mm have been considered. The results are shown for the 10% ethanol based ice slurry with particle diameter of 0.25 mm. It is seen that, the value of critical velocity increases with increase in pipe diameter. The pipe of bigger diameter has bigger cross-section area that increases the rate of moving bed formation due to which value of critical velocity is increased.

5. Conclusions

- Critical velocity increases with increase in initial particle concentration. The value of critical velocity for 0.4 particle diameter increases from 0.16 m/s to 0.30 m/s with the increase in from 10% to 30%.
- Critical velocity also depends on particle diameter and it increases from 0.4 m/s to 0.16 m/s with increase in particle size from 0.1 mm to 0.4 mm at 10 % initial particle concentration.
- The value of critical velocity increases with decrease of ethanol concentration. At lower ethanol concentration in ice slurry, the density difference between the phases becomes large and viscosity of liquid phase also decreases due to which critical velocity increases.
- The value of critical velocity increases with pipe diameter. The large diameter has bigger cross-section area that increases the rate of moving bed formation due to which value of critical velocity is increased. Therefore the value of pipe diameter should be small to avoid chances of formation of moving bed. The value of critical velocity for ice slurry at 10% initial particle concentration and 0.25 particle size has been determined in 23, 50 and 100 mm diameters pipe and it is found to be 0.6, 0.12 and 0.19 m/s, respectively.

References

[1] Desideri, U., Proietti, S. and Sdringola, P., 2009. Solar-powered cooling systems: Technical and economic analysis on industrial refrigeration and air-conditioning applications. Applied Energy, 86(9), pp.1376-1386.

[2] Wang K., Eisele M., Hwang Y., and Radermacher R., 2010. Review of secondary loop refrigeration systems. International Journal of Refrigeration, 33(2), pp.212-234.

[3] Melinder, Å., 2010. Properties and other aspects of aqueous solutions used for single phase and ice slurry applications. International journal of refrigeration, 33(8), pp.1506-1512.

[4] Delgado M., Lázaro A., Mazo J. and Zalba B., 2012, Review on phase change material emulsions and microencapsulated phase change material slurries: materials, heat transfer studies and applications. Renewable and Sustainable Energy Reviews, 16(1), pp.253-273.

[5] Delgado M., Lázaro A., Peñalosa C., Mazo J. and Zalba B., 2013, Analysis of the physical stability of PCM slurries. International Journal of Refrigeration, 36(6), pp.1648-1656.

[6] Chandel S.S. and Agarwal T., 2017. Review of current state of research on energy storage, toxicity, health hazards and commercialization of phase changing materials. Renewable and Sustainable Energy Reviews, 67, pp.581-596.

[7] Tay N.H.S., Liu M., Belusko M. and Bruno F., 2017. Review on transportable phase change material in thermal energy storage systems. Renewable and Sustainable Energy Reviews, 75, pp.264-277.

[8] Rawat, K.S. and Pratihar, A.K., 2018, February. Numerical Investigation of Ice Slurry Flow in a Horizontal Pipe. In IOP Conference Series: Materials Science and Engineering (Vol. 310, No. 1, p. 012095). IOP Publishing.

[9] Wang J., Wang S., Zhang T. and Liang Y., 2013, Numerical investigation of ice slurry isothermal flow in various pipes. International Journal of Refrigeration, 36(1), pp.70-80.

[10] Kitanovski A., Vuarnoz D., Ata-Caesar D., Egolf P.W., Hansen T.M., and Doetsch C., 2005, The fluid dynamics of ice slurry. International Journal of Refrigeration, 28(1), pp.37-50.

[11] Tian Q., He G., Wang H. and Cai D., 2014, Simulation on transportation safety of ice slurry in ice cooling system of buildings. Energy and Buildings, 72, pp.262-270.

[12] Lun C. K. K., Savage S. B., Jeffrey D. J., Chepurniy, N., 1984. Kinetic theories for granular flow: inelastic particles in Couette flow and slightly inelastic particles in a general flowfield. Journal of fluid mechanics, 140, 223-256.

[13] Zhang P. and Shi X.J., 2015, Thermo-fluidic characteristics of ice slurry in horizontal circular pipes. International Journal of Heat and Mass Transfer, 89, pp.950-963.

[14] Gidaspow D., 1994, Multiphase Flow and Fluidization. Academic Press, Boston, MA.

[15] Gidaspow D., Bezburuah R., and Ding J., 1992, "Hydrodynamics of Circulating Fluidized Beds, Kinetic Theory Approach". In Fluidization VII, Proceedings of the 7th Engineering Foundation Conference on Fluidization. 75–82.

[16] Kauffeld M., Kawaji M. and Egolf W.P., 2005, Handbook on ice slurries. International Institute of Refrigeration, Paris, 359.

[17] Rawat, K.S. and Pratihar, A.K., 2018. Thermo-Physical Properties of Aqueous Solution of Ice Slurry. International Journal of Pure and Applied Mathematics, 119(17), pp.1493-1497.

IOP Conf. Series: Materials Science and Engineering **802** (2020) 012003 doi:10.1088/1757-899X/802/1/012003

Evaluation of Influence of Finish line Design on Marginal Discrepancy of All-ceramics Lithium disilicate Crown restorations using µ-CT

Yajvinder[1], Vishal Gulati[2], Ashish Agrawal[3], Balwant Singh[4]

1,2, Guru Jambheswar Unversity of science & Technology, India
3,4, Techchnical Physics Department, BARC, India

Jnvd1994yaj@gmail.com

Abstract. Purpose of this study was to evaluate influence of finish line design on marginal discrepancies of lithium disilicate crown restoration fabricated with CAD/CAM. 24 crown restorations (n=8 each group) were designed with three different finish lines: Shoulder finish line (SFLC), Round Shoulder finish line (RSFLC) and Deep Chamfer finish line. All dies were prepared with additive manufacturing whereas crowns were fabricated with CAD/CAM using these preparations. After fixation, restorations were scanned using SRµCT and obtained reconstructed 2D slices were used to measure marginal discrepancy (MD) and absolute marginal discrepancy (AMD) at four sites (buccal, lingual, mesial and distal) of each specimen with the help of Image J software. Finally, recorded data was analyzed by one way ANOVA and Tukey's test. The mean marginal discrepancies (MD±SD) for three groups were as follow: SFLC (91.6 ±3.5), RSFLC (72.8 ±3.9) and DCFLC (65.9 ±5.2) whereas Mean absolute marginal discrepancies (AMD±SD) were as follow: SFLC (104.4 ±6.8), RSFLC (85.5 ±3.5) and DCFLC (78.7 ±5.3). According to results, marginal discrepancies as well as absolute marginal discrepancies of SFLC group were significantly higher than those of RSFLC and DCFLC groups. After all, MDs and AMDs of all three groups were below clinical acceptable value. For all-ceramics restorations, marginal discrepancies of lithium dislicate crown restorations were below clinical acceptable range and restorations designed with RSFLC and DCFLC demonstrated superior results compared to preparations designed with SFLC.

1. Introduction

Among the various factors responsible for the longevity of a complete crown restoration, marginal discrepancy plays a vital role [1,2,3]. In general, marginal discrepancy is explained in terms of marginal discrepancy or gap, which can be defined as the vertical distance from the finish line of prepared tooth to the cervical margin of the restorations [4]. Poor marginal discrepancy creates high plaque accumulation and exposes the cemented region to oral environment, which increases micro-leakage and leads to periodontal diseases. Finally restoration losses its mechanical stability and failures occur [5,6].

The past studies acknowledged that cast restorations have shown better marginal discrepancy (even marginal gap below 10 µm have been reported in some studies for high-palladium-content alloys and gold-platinum-palladium alloys) than base metal, metal ceramic and all ceramic CAD/CAM fabricated restorations [7,8], however, in dentistry, applications of casting process is limited to few metals which are high in cost and do not satisfied the esthetic issues for patients. In other hand, due to advancement in dental CAD/CAM various biomaterials such as ceramics are in practice for dental restorations [9,10,11,12]. In case of dental ceramic materials, lithium disilicate glass ceramic has been used for fabrication of anterior or posterior crowns, inlays, onlays, veneers since it was first introduced in dentistry by Ivoclar Vivadent in 1990 [13,14,15]. Furthermore, pre-crystallized lithium disilicate blocks can be easily milled without excessive diamond bur wear or damage to the restoration and finally, restoration posses fine graine size, natural tooth shade and superior mechanical properties [15]. As far as maximum marginal discrepancy, no pre established acceptable range of marginal discrepancy and researchers suggested it according to their convenience in initial studies. Recently, several studies recommended that marginal discrepancies up to 120 μm are considered clinically acceptable. As CAD/CAM process reduces the possibility of human error and the inaccuracy inherent in different restorative materials, 100 μm is the maximum acceptable marginal discrepancy for CAD/CAM restorations [16,17].

In past three decades, various studies investigated the factors those influence the marginal discrepancy of all-ceramics crown restoration. Many of these studies reported that factors such as finish line design, veneering firing cycles, angle of preparation, cement space, type of cementation, and material used reportedly influence the marginal discrepancy of complete crown restoration [18,19,20,21,22,23].

Interstingly, for conventional full crown restorations, effect of finish line design on marginal discrepancy denied by several investigators in their reports but studies on CAD/CAM fabricated metal-ceramic and all-ceramics crown restorations reported contradictory results. Some studies showed that no significant difference between two finish line design whether other studies recommended that finish line design influence the marginal discrepancy [18,20,24,25,26,27]. A study by Akbar Hussain et al [26] based on Cerec3 system compared two different finish line designs reported that no significant effect of finish line design on marginal discrepancy whereas in two different studies by same Raul Euan et al [16,27] finish line design significantly influences the marginal discrepancies. However, due to heterogeneity in experimental environment, direct comparison of these studies was impossible. Although, heavy or deep chamfer and round shoulder finish line designs have been recommended for all-ceramic crown restorations [4,18,28,29,30] but adequate literature is not available. So investigation of influence of finish line design on marginal discrepancy of all-ceramics restorations is needful.

However, in vivo studies, marginal discrepancy can be measured by various methods including stereomicroscope, scanning electron microscopy, optical microscopy, optical comparator screen, and computed tomography [31,32,33,34] whereas various techniques including silicon replica, resin replica, and marginal interval measurement by direct sectioning of a specimen can be used for in vitro studies. Among all, method for evaluation of marginal and internal discrepancy of restorations by μ-CT is noninvasive and non destructive [35]. However, high quality 2D and 3D images from any angle and position can be obtained, and very thin sections may also be imaged safely from close proximity[36,37], but formation of radiation artifacts caused by the differences of radiation absorption coefficients between different materials during cement space volume measurement is the major disadvantage of this method. [38]

The purpose of this study was to evaluate the marginal discrepancies of CAD/CAM fabricated all-ceramics lithium disilicate crown restorations having three different finish line designs (shoulder, round shoulder and deep chamfer) using μ-CT. All three groups were evaluated to test the following hypotheses: null hypothesis that no difference would be found among the marginal discrepancies measured in different locations for these finish line designs; and alternative hypothesis that finish line influences the marginal discrepancies of the all-ceramics crown restorations.

2. Materials and Methods

An extracted human lower left second premolar tooth with no carries or anatomical defect was used in this study. Tooth was obtained following the guidelines of local human research ethics committee and stored in 10% formaldehyde solution until it was used for study. The extracted tooth was scanned using 3D blue light scanner (Shining3D, DS100) and therefore obtained STL file was then sent and used for CAD modeling using Solid works CAD software where three CAD models were designed. These CAD models were classified by three different finish line configurations i.e. Shoulder Finish Line Configuration (SFLC), Round Shoulder Finish Line Configuration (RSFLC), and Deep Chamfer Finish Line Configuration (DCFLC).

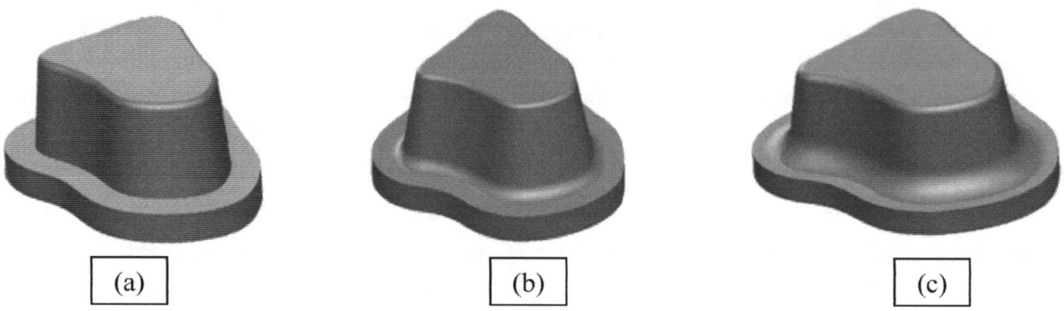

Figure 1: Preparations CAD models with (a) Shoulder finish line, (b) Round Shoulder finish line and (c) Deep Chamfer finish line.

categorize the three groups of preparations but all other parameters were common and had following characteristics: approx. 12 degree convergence angle, 1.5 to 2 mm anatomic occlusal reduction, axial reduction of 1.0 to 1.2 mm, and a finish line located 0.5 mm above the cementoenamel base. For first group, shoulder finish line was designed whereas round shoulder finish line and chamfer finish line were designed for second group and third group respectively.

In this study eight specimens were included from each group and all 24 master dies (or preparations) of specimens were fabricated by Rapid Prototyping Technique (RPT) using clear resin material (SLA FOAM LAB). Then all fabricated master dies sprayed with white color and scanned with 3D blue light scanner (Shining3D, DS100). The obtained scanned files were then sent to crown designing unit where crowns were designed for each master die with EXOCAD dental CAD software. All the crowns were designed under following manufacturer's instruction: 0.5 mm wall thickness, a 0.35 mm reinforcement of the restoration edge, and 40 micron cement space.

In milling unit, two different size burs (2.5 mm & 1 mm) were used to manufacture each of the all 24 crowns from lithium disilicate block (IPS e.max CAD). All the manufactured crowns were operated under 850 porcelain firing cycles for 20 min in a single shift and no additional finishing or polishing was carried. Therefore, complete specimens were prepared by fixing of each crown on its respective master die with ionomiser cement solution under a constant load of 20 N for 20 min.

Synchrotron Radiation (SR) x-ray microtomogramphy (SRμ-CT) was used to obtain high quality images of all the assembled specimens which were analyzed to evaluate the marginal discrepancies. Therefore, each specimen was fixed on experimental table and positioned perpendicular to the x-ray beam for scanning process. In this scanning process, the SR-Micro CT was operated at 25KV energy and between 100 to 165mA x-ray beam line current. Total 901 images were acquired for every specimen whereas each image capturing time was 6 to 10 seconds.

All acquired projections were reconstructed and translated into cross-section slices with the help of image processing software and after that image analysis software (ImageJ, 1.51p22) was used to analyze the reconstructed cross-sectional images. For all 24 specimens, marginal discrepancies were measured in both buccolingual and mesiodistal directions hence in this course five slices for buccolingual direction (slice no. 1,5,10, 890 & 895 of each specimen) and five slices for mesiodistal direction (Slice no. 215,220, 225, 230 & 235 of each specimen) were selected and evaluated for each specimen. For each selected slice two regions (buccal and lingual areas on buccolingual slices whereas mesial and distal areas on mesiodistal slices) were marked and evaluated for two measurements: MD, AMD [fig] thus each specimen had total 40 measurements. Finally, one way ANOVA technique was implemented using the statistical software (IBM

SPSS Statistics 20) to perform statistical analyses. Tukey's (HSD) test was used for homogeneous variance groups whereas for non-homogeneous groups Tamhane T2 test was performed. (α =0.05)

3. Results

For each of three SFLC, RSFLC and DCFLC groups including 8 crowns/group and 4 sites/ crown, total 80 images were analyzed. Descriptive analyses of the AMDs and MDs statistics is given in Table 2 and Table 3 respectively. Mesial site of SFLC group attains highest AMD (122.43μm) that was significantly higher than minimum AMD (65.23μm) was found at buccal site of DCFLC group (p<0.01). On average, AMDs of SFLC group were significantly higher than those for RSFLC and DCFLC groups (p<0.01). In case of MDs, the highest MD (104.02μm) which measured at mesial site of SFLC group was significantly higher than minimum MD recorded in DCFLC group for buccal site (p<0.01). Overall, MDs of SFLC group were seen significantly higher than those of the RSFLC and DCFLC groups (p<0.01).

For the SFLC group, highest AMD (122.43μm) was seen at mesial site that was also significantly higher than lowest AMD (90.43μm) recorded at lingual site (p<0.01). However, rather than AMDs significantly higher than those of lingual site (p<0.05) there was no significant difference between AMDs of all other sites. Maximum value of MD (104.02μm) was seen at mesial site that was also significantly higher than minimum MD value (82.73μm) at buccal site. In average, MDs at mesial site were significantly higher than those of buccal and distal sites (p<0.01) as well as of lingual site (P<0.05).

Table 1: Mean and standard deviation values of MD (μm)

Group/Site	**Buccal Mean ±SD (a)**	**Lingual Mean ±SD (b)**	**Mesial Mean ±SD (c)**	**Distal Mean ±SD (d)**	**Mean ±SD ***	***P*+**
S FLC (A)	86.0 ±3.2	91.5 ±4.2	99.1 ±3.5 ^{a}	89.9 ±3.7 ^{c}	91.6 ±3.5	**<0.01**
RS FLC (B)	66.3 ±2.4^{A}	70.3 ±8.4 ^{a,A}	72.4 ±3.8 ^{a,b,A}	72.2 ±2.5 ^{a,c,A}	72.8 ±3.9 ^{A}	**<0.01**
DC FLC (C)	70.8 ±4.9 ^{A}	65.1 ±4.6 ^{A}	71.1 ±7.2 ^{a,A,B}	67.3 ±4.6 ^{a,A}	65.9 ±5.2 ^{A}	**<0.01**
***P*~**	**<0.01**	**<0.01**	**<0.01**	**<0.01**	**<0.01**	

MD, Marginal Discrepancy; SD, Standard Deviation; S, Shoulder; RS, Round Shoulder; DC, Deep Chamfer; FLC, Finish Line Configuration; Mean values followed by different lowercase letters in same row show statistical differences (*p* <0.05). Mean values followed by different uppercase letters in same column show statistical differences (*p* <0.05). * Averagevalue (Mean ±SD) of buccal,lingual, mesial and distal. + Repeated measures analysis of variance. ~ One way ANOVA

Table 2: Mean and standard deviation values of AMD (μm)

Group/Site	**Buccal Mean ±SD (a)**	**Lingual Mean ±SD (b)**	**Mesial Mean ±SD (c)**	**Distal Mean ±SD (d)**	**Mean ±SD ***	***P*+**
S FLC (A)	96.9 ±6.3	104.2 ±8.8	114.7 ±5.2 ^{a,b}	104.4 ±6.8 ^{c}	104.4 ±6.8	**<0.01**
RS FLC (B)	75.2 ±3.6	83.2 ±3.1 ^{A}	98.4 ±5.5 ^{a,b,A}	85.0 ±3.9 ^{c,A}	85.5 ±3.5 ^{A}	**<0.01**
DC FLC (C)	70.0 ±3.2	76.0 ±6.0 ^{A}	86.0 ±6.2 ^{a,A}	82.8 ±5.3 ^{A}	78.7 ±5.3 ^{A}	**<0.01**
***P*~**	**<0.01**	**<0.01**	**<0.01**	**<0.01**	**<0.01**	

AMD, Marginal Discrepancy; SD, Standard Deviation; S, Shoulder; RS, Round Shoulder; DC, Deep Chamfer; FLC, Finish Line Configuration; Mean values followed by different lowercase letters in same row show statistical differences (p <0.05). Mean values followed by different uppercase letters in same column show statistical differences (p <0.05). * Averagevalue (Mean ±SD) of buccal,lingual, mesial and distal. + Repeated measures analysis of variance. ~ One way ANOVA.

For RSFLC group, highest AMD (106.54μm) was found at mesial site and noticeably that was significantly higher than lowest AMD (71.47μm) located at buccal site (p<0.01). AMDs of mesial site were significantly higher than those of buccal, lingual and distal sites (p<0.01). Top most MD value (87.49μm) was also found for mesial site whereas with significant difference (p<0.01), lower most MD value (56.04μm) was seen at buccal site. MDs of mesial site were significantly greater than MDs of buccal, lingual and distal sites (p<0.01).

For DCFLC group, In AMD values highest (94.77μm) were seen for mesial and distal sites that were significantly higher than lowest value recorded for buccal site (p<0.01). Overall, AMDs seen in mesial site were significantly higher than AMDs of buccal site (p<0.05). In case of MDs, highest (77.62μm) and lowest (52.11μm) values had significant difference and were located at mesial site and buccal site respectively (p<0.01). In comparison, mesial site's MDs were significantly higher than that of buccal site (p<0.01).

4. Discussion

Aim of this study was to evaluate influence of the different finish line configurations of preparation on the marginal discrepancy of all-ceramics crown restoration using Micro CT. The null hypothesis that stated there was no significant difference in tested group should be rejected on the basis of present study.

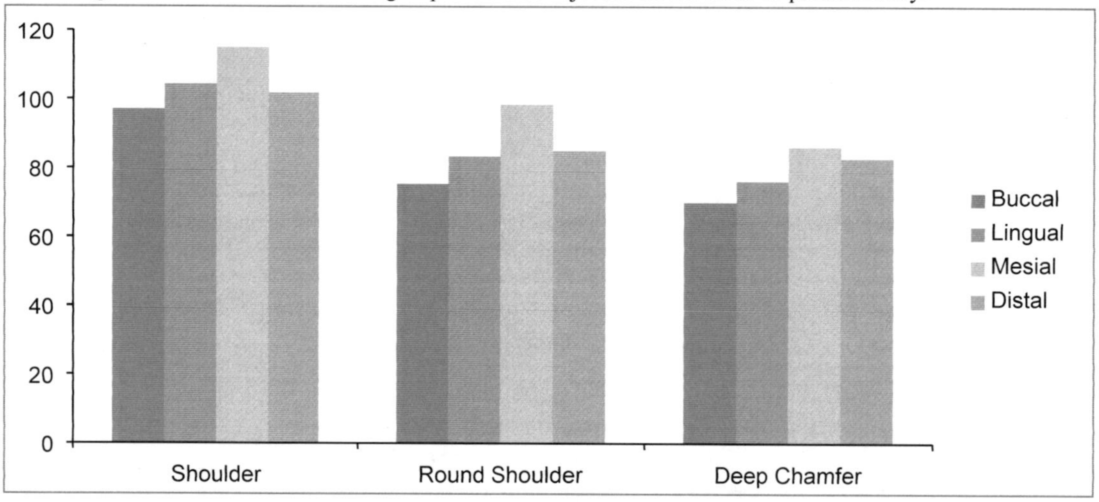

Figure 2: Comparison of AMD mean of three groups at buccal, lingual, mesial and distal locations.

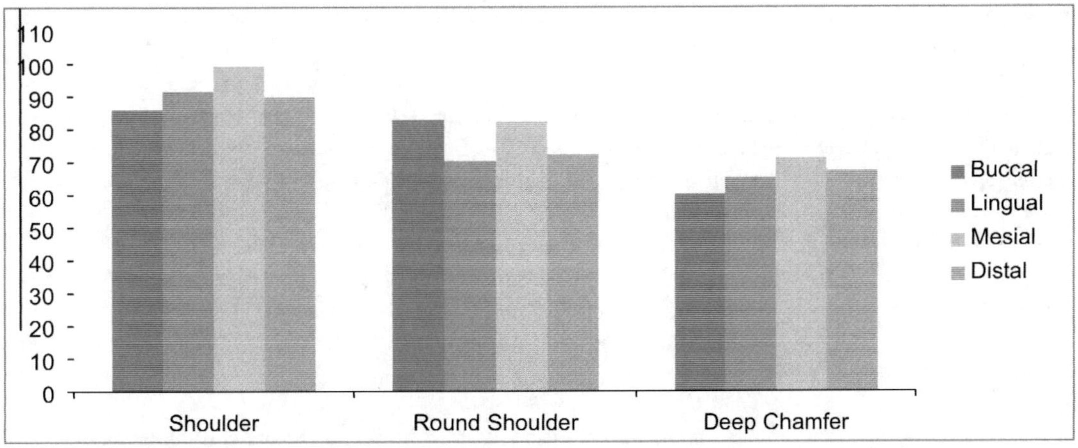

Figure 3: Comparison of MD mean of three groups at buccal, lingual, mesial and distal locations.

Minimal marginal discrepancy remains essential factor for clinical success of all-ceramics crown restorations. However, marginal discrepancy influences by various manufacturing process parameters but finish line configuration of preparation is one of the primary stage design parameter. In addition, method used for measurement of marginal discrepancy is another whole process influencing factor. In the literature, in order to evaluate marginal discrepancy, various destructive and nondestructive, and quantitative and qualitative measurement methods have been applied. Micro CT, a non destructive evaluation method has capability of 2D and 3D investigations of the marginal and internal fit at multiple sites and directions within a few micrometers [39,40,41]. Major disadvantages of this technique are the low power of distinction in inadequate radiographic contrast cases and radiographic artifacts [42]. In the present study, lithium disilicate crown restorations are introduced for investigation of influence of different finish line configurations on marginal discrepancy of all-ceramics crown restorations. To apart the disadvantages of Micro CT measurement approach, Synchrotron X-Ray Micro CT was used. In this study, Image J software implemented to investigate the marginal discrepancy at 2D reconstructed slices.

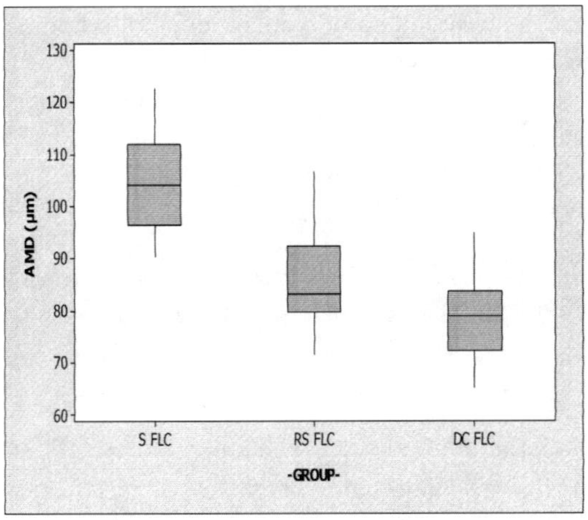

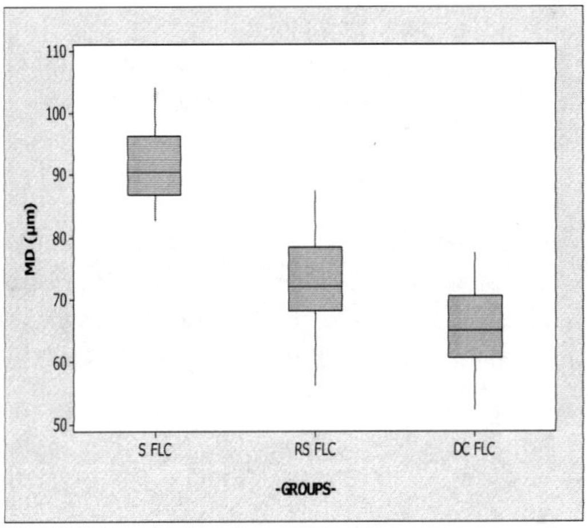

Figure 4: MD Mean comparison of groups. of groups.

Figure 5: AMD Mean comparison

However, many researchers have criticized the marginal and internal discrepancies of CAD/CAM fabricated crown restorations [42,43]. Deoggu et al [44] investigated restoration prepared with shoulder finish line. The mean marginal discrepancies seen in this study were between 200.2 to 253.3µm. But recent studies reported better marginal discrepancy results for crown restoration fabricated with CAD/CAM. Euan et al [27] found marginal discrepancies between 17.2 to 72.8µm for group of crown restorations with chamfer finish line whereas those for round shoulder finish line crown restoration group were between 13.9 to 63.1µm. The differences could be explained by restoration fabrication process and discrepancy measurement method used in these studies. Moreover, no clinical guidelines of acceptable marginal discrepancy value for all-ceramics crown restorations fabricated with CAD/CAM in the literature but the maximum clinical acceptable marginal discrepancy value 165µm suggested by some authors [45,46]. In later studies below 150µm considered acceptable value [47,48]. A clinical study by Maclean and Fraunhofer over 5 years included 1000 restorations, recommended 120µm as upmost marginal discrepancy value [17].

Although, various studies investigated the marginal discrepancy of lithium disilicate crown restorations. In a research by Kim et al [35] three groups of mandibular first molar classified by different impression techniques and manufactured with CAD/CAM were investigated with Micro CT and found that the marginal discrepancies MDs of three lithium dislicate crown restoration groups were between 129µm and 207µm whereas the maximum mean absolute marginal discrepancy AMD seen in this study was 155.5µm which were significantly higher than those of reported in another study by Mously et al [42] In addition, Reich et al [11] reported marginal discrepancy about 101µm whereas Yildrim et al [38] found it 129µm in their respective studies for IPS crown restorations. In the present vitro study including all three groups, results for maximum mean marginal discrepancy MD (99.6µm) and absolute marginal discrepancy AMD (104.4µm) were below clinical acceptable value (120µm). However, marginal discrepancies of both DCFLC group and RSFLC group had no significant differences but these were significantly lower than those of SFLC group. The reasons of this difference may be (a) Preparations with shoulder finish line had more depth than those of preparation with chamfer finish line and round shoulder finish line. (b) Due round angle between gingival and axial surfaces, the preparations with chamfer finish line provided more accurate seat for the crown whereas preparations with shoulder finish line had slight round angle that lead to partial seat of crown and enlarged the vertical gap. The other existing causes may be limited [49,50,51].

Nevertheless, other than finish line design, various other design parameters i.e. convergence angle, cement space; fabrication process parameters i.e. impression technique, scanner, milling process, porcelain firing; as well as discrepancy measurement technique influence the marginal and internal discrepancies of all-ceramics crown restoration. In this study, to provide the homogeneity in complete fabrication process and experimental conditions, all the preparations and crowns were designed using common parameters while fabricated with same manufacturing setup. However, during scanning process the beam current of Micro CT was not constant but tube power was sufficient to achieve necessary image resolution.

5. Conclusion

For all-ceramics restorations, marginal discrepancies of lithium dislicate crown restorations were below clinical acceptable range and restorations designed with RSFLC and DCFLC demonstrated superior results compared to preparations designed with SFLC.

6. References

[1] A Bottino MA, Valandro LF, Buso L, Ozcan M. The influence of cervical finish line, internal relief, and cement type on the cervical adaptation of metal crowns. Quintessence Int 2007;38:425-32.
[2] Ayad MF. Effects of tooth preparation burs and luting cement types on the marginal fit of extra coronal restorations. J Prosthodont 2009;18:145-51.
[3] Limkangwalmongkol P, Kee E, Chiche GJ, Blatz MB. Comparison of marginal fit between all-porcelain margin versus alumina-supported margin on Procera alumina crowns. J Prosthodont 2009;18:162-6.

[4] Holmes JR, Bayne SC, Holland GA, Sulik WD. Considerations in measurement of marginal fit. J Prosthet Dent 1989;62:405-8.
[5] Beschnidt SM, Strub JR. Evaluation of the marginal accuracy of different all ceramic crown systems after simulation in the artificial mouth. J Oral Rehabil 1999;26:582-93.
[6] Jacobs MS, Windeler AS. An investigation of dental luting cement solubility as a function of the marginal gap. J Prosthet Dent 1991;65:436-42
[7] A Byrne G, Goodacre CJ, Dykema R, Moore BK. Casting accuracy of high-palladium alloys. J Prosthet Dent 1986;55:297-301.
[8] Syu JZ, Byrne G, Laub LW, Land MF. Influence of finish-line geometry on the fit of crowns. Int J Prosthodont 1993;6:25-30.
[9] Takeichi T, Katsoulis J, Blatz MB. Clinical outcome of single porcelain-fusedto- zirconium dioxide crowns: a systematic review. J Prosthet Dent 2013;110: 455-61.Conrad HJ, Seong WJ, Pesun IJ. Current ceramic materials and systems with clinical recommendations: a systematic review. J Prosthet Dent 2007;98: 389-404.
[10] Syrek A, Reich G, Ranftl D, Klein C, Cerny B, Brodesser J. Clinical evaluation of all-ceramic crowns fabricated from intraoral digital impressions based on the principle of active wavefront sampling. J Dent 2010;38:553-9.
[11] Kokubo Y, Nagayama Y, Tsumita M, Ohkubo C, Fukushima S, Vult von Steyern P. Clinical marginal and internal gaps of In-Ceram crowns fabricated using the GN-I system. J Oral Rehabil 2005;32:753-8.
[12] Dong JK, Luthy H, Wohlwend A, Scharer P. Heat-pressed ceramics: technology and strength. Int J Prosthodont 1992; 5(1):9-16.
[13] El-Mowafy O, Brochu JF. Longevity and clinical performance of IPS-empress ceramic restorations--a literature review. J Can Dent Assoc. 2002 Apr;68(4):233-7.
[14] Tysowsky G. the science behind lithium disilicate: today's surprisingly versatile, esthetic and durable metal-free alternative. Oral Health J 2009;March:93-97.
[15] Euán R, Figueras-Álvarez O, Cabratosa- Termes J, Brufau-de Barbera M, Gomes- Azevedo S. Marginal adaptation of zirconium dioxide copings: Influence of the CAD/CAM system and the finish line design. J Prosthodont 2014.
[16] McLean JW, von Fraunhofer JA. The estimation of cement film thickness by an in vivo technique. Br Dent J 1971;131:107-11.
[17] Ferreira A, Oliveira F, Bottino MA: Vertical marginal discrepancy of ceramic copings with different ceramic materials, finish lines, and luting agents: an in vitro evaluation. J Prosthet Dent 2004;92:250-257
[18] Vigolo P, Fonzi F: An in vitro evaluation of fit of zirconium-oxide based ceramic four-unit fixed partial dentures, generated with three different CAD/CAM systems, before and after porcelain firing cycles and after glaze cycles. J Prosthodont 2008;17:621-626
[19] Komine F, Takayuki I, Kobayashi K, et al: Marginal and internal adaptation of zirconium dioxide ceramic copings and crowns with different finish line designs. Dent Mater J 2007;26: 659-664
[20] Wolfart S, Wegner SM, Al-Halabi A, et al: Clinical evaluation of marginal fit of a new experimental all-ceramic system before and after cementation. Int J Prosthodont 2003;16:587-592
[21] Cho L, Choi J, Yi YJ, et al: Effect of finish line variants on marginal accuracy and fracture strength of ceramic optimized polymer/fiber-reinforced composite crowns. J Prosthet Dent 2004;91:554-560
[22] Gavelis JR, Morency JD, Riley ED, et al: The effect of various finish line preparations on the marginal seal and oclusal seat of full crown preparations. J Prosthet Dent 1981;45: 138-145
[23] Suarez MJ, Gonzalez de Villaumbrosia P, Pradies G, Lozano JF. Comparison of the marginal fit of Procera AllCeram crowns with two finish lines. Int J Prosthodont 2003;16:229-32.
[24] Alghazzawi TF, Liu PR, Essig ME. The effect of different fabrication steps on the marginal adaptation of two types of glassinfiltrated ceramic crown copings fabricated by CAD/CAM technology. J Prosthodont 2012;21:167-72.
[25] Akbar JH, Petrie CS, Walker MP, Williams K, Eick JD. Marginal adaptation of Cerec 3 CAD/CAM Composite Crowns using Two different Finish Line Preparation designs. J Prosthet Dent 2006;15(3): 155-63.
[26] Euan R, Figueras O, Cabratosa J, Brufau M , Gomes S Comparison of the Marginal Adaptation of Zirconium Dioxide Crowns in Preparations with Two Different Finish Lines. Journal of Prosthodontics **00** (2012) 1–5.

[27] Karatasli O, Kursoglu P, Çapa N, Kazazoglu E. Comparison of the marginal fit of different coping materials and designs produced by computer aided manufacturing systems. Dent Mater J 2011;30(1): 97-102.
[28] Yeo IS, Yang JH, Lee JB. In vitro marginal fit of three all-ceramic crown systems. J Prosthet Dent 2003;90(5):459-64.
[29] Goldin EB, Boyd NW 3rd, Goldstein GR, Hittelmann EL, Thompson VP. Marginal fit of leucite-glass pressable ceramic restorations and ceramic-pressed-to-metal restorations. J Proshtet Dent 2005;93: 143-7.
[30] Ng J, Ruse D, Wyatt C. A comparison of the marginal fit of crowns fabricated with digital and conventional methods. J Prosthet Dent 2014; 112: 555-60.
[31] An S, Kim S, Choi H, Lee JH, Moon HS. Evaluating the marginal fit of zirconia copings with digital impressions with an intraoral digital scanner. J Prosthet Dent 2014;112:1171-5. 21.
[32] Xu D, Xiang N, Wei B. The marginal fit of selective laser melting-fabricated metal crowns: An in vitro study. J Prosthet Dent 2014;112: 1437-40.
[33] Shembesh M, Ali A, Finkelman M, Weber HP, Zandparsa R. An in vitro comparison of the marginal adaptation accuracy of CAD-CAM restoration using different impression systems. J Prosthodont 2016
[34] Kim JH, Jeong JH, Lee JH, Cho HW. Fit of lithium disilicate crowns fabricated from conventional and digital impressions assessed with micro-CT. J Prosthet Dent 2016;116:551-7.
[35] Contrepois M, Soenen A, Bartala M, Laviole O. Marginal adaptation of ceramic crowns: a systematic review. J Prosthet Dent 2013;110: 447-54.
[36] Demir N, Ozturk AN, Malkoc MA. Evaluation of the marginal fit of full ceramic crowns by the microcomputed tomography (micro-CT) technique. Eur J Dent 2014;8: 437-44.
[37] Yildirim G, Uzun HI, Keles A. Evaluation of marginal and internal adaptation of hybrid and nanoceramic systems with microcomputed tomography: An in vitro study. J Prosthet Dent 2016
[38] Pelekanos S, Koumanou M, Koutayas SO, Zinelis S, & Eliades G. Micro-CT evaluation of the marginal fit of different In-Ceram alumina copings. European Journal of Esthetic Dentistry 2009; 4(3); 278-292. 17.
[39] Seo D, Yi Y, & Roh B. The effect of preparation designs on the marginal and internal gaps in Cerec3 partial ceramic crowns. Journal of Dentistry 2009; 37(5); 374-282.
[40] Borba M, Cesar PF, Griggs JA, & Della Bona A. Adaptation of all-ceramic fixed partial dentures Dental Materials 2011; 27(11); 1119-1126.
[41] Mously HA, Finkelman M, Zandparsa R, & Hirayama H. Marginal and internal adaptation of ceramic crown restorations fabricated with CAD/CAM technology and the heat-press technique Journal of Prosthetic Dentistry 2014; 112(2); 249-256.
[42] Bayne SC, Heymann HO. CAD/CAM in dentistry: Present and future applications. Quintessence International (Current Concepts) 1996; 27(6); 431-433.
[43] Deoggyu S, Youngah Y, Byoungduck R. The effect of preparation designs on the marginal and internal gaps in Cerec3 partial ceramic crowns. J Dent 2009;37: 374-82.
[44] Schaerer P, Sato T, Wohlwend A. A comparison of the marginal fit of three cast ceramic crown systems. J Prosthet Dent 1988;59:534-42.
[45] Sulaiman F, Chai J, Jameson LM, Wozniak WT. A comparison of the marginal fit of In-Ceram, IPS Empress, and Procera crowns. Int J Prosthodont 1997;10(5):478-84.
[46] Fransson B, Oilo G, Gjeitanger R. The fit of metal-ceramic crowns: a clinical study. Dent Mater 1985;1:197-9.
[47] Boening KW, Walter MH, Reppel PD. Non-cast titanium restorations in fixed prosthodontics. J Oral Rehabil 1992;19:281-7.
[48] Lombardas P, Carbunaru A, McAlarney ME, Toothaker RW. Dimentional accuracy of castings produced with ringless and metal ring investment systems. J Prosthet Dent 2000;84(1):27-31.
[49] Kyu-Bok, Charn-Woon, Kyo-Han and Tae-Yub. Marginal and internal fit of all-ceramic crowns fabricated with two different CAD/CAM systems. Dent Mater J 2008;27(3):422-6.
[50] Park SH, Lee KB. A comparison of the fidelity between various cores fabricated with CAD/CAM system. J Kor Acad Prosthodont 2008;43(3): 269-78.
[51] Al-Zubaidi Z, Al-Shamma A. The Effect of Different Finishing Lines on the Marginal Fitness of Full Contour Zirconia and Glass Ceramic CAD/CAM Crowns (An in-vitro study). JDMT 2015;4(3): 127-36.

IOP Conf. Series: Materials Science and Engineering **802** (2020) 012004 doi:10.1088/1757-899X/802/1/012004

The effect of alumina on mechanical behavior of Al 6064 alloy

Abhishek Kumar[1,2,3], Paras Kumar[1], R C Singh[1]
[1]Department of Mechanical Engineering, Delhi Technological University, Bawana Road, Delhi-110042 India
[2] Department of Mechanical Engineering, IILM College of Engineering & Technology, Greater Noida, Uttar Pradesh-201306 India

[3]Corresponding author email: 2003.abhi@gmail.com

Abstract-
Need in the development of new materials with excellent properties are growing day by day. Composite materials are the simple and better way to accomplish materials with superior properties. Aluminium metal matrix composites due to their high strength to weight ratio, low cost, improved electrical conductivity and good tribological properties are broadly produced and used in various engineering applications in the area of automobile, aerospace and military engineering. In the present paper an attempt has been made to fabricate MMC of an alloy of aluminium 6064, reinforced with alumina particles through stir casting with vortex technique. Three aluminium alloy composites with alumina (2.5-5-7.5 wt. %) were fabricated. Different mechanical tests for hardness, tensile strength and toughness were carried out for the samples. Addition of alumina to aluminium alloy shows an increase in mechanical properties of aluminium metal matrix composite. The results showed that the stir casting could produce uniformly distributed reinforced particles of Al_2O_3 in Al 6064 alloy.

1.Introduction

Metal matrix composites usually consists of aluminum, magnesium or its alloys reinforced with particulate or fibers of a ceramic material, such as alumina, silicon carbide, TiC, TiO_2, B_4C etc. On comparison with unreinforced metals, MMCs offer better strength to weight ratio, improved resistance to wear and abrasion, higher operating temperature, and greater electrical conductivity. However, MMCs also have some disadvantages such as high coefficient of thermal expansion, higher stiffness. Most among these are the higher cost of fabrication for high-performance MMCs, lower ductility and toughness. At present aluminium MMCs have two extreme types. First abide of very high performance composites reinforced with costly continuous fibers and requiring expensive processing methods. Second abide of relatively low-cost and low-performance composites reinforced with relatively less costly particulate or fibers. The cost of the first type is too high for any but military or space applications, whereas the cost/benefit advantages of the second type over unreinforced metal alloys remain in doubt [1]. Many factors which influence the effect of viscosity during Al–SiC MMC production. Processing periods (up to 65 min), stirring speeds (50–500 rpm), and reinforcement sizes (13–100 μm) for two different viscosity levels (1 and 300 mPa-s) were investigated [2]. Particulate metal matrix composites (PMMCs) have attracted interest for application in numerous fields [3]. The wear rate of spray formed composites is significantly lower than the base alloy and stir cast composite under same sliding condition [4]. AA6063 aluminium matrix composite billets reinforced with SiC particles (Al/SiCp) were prepared by stir casting method for extrusion process. The results show that the die wear is affected differently by the SiC particle size [5]. Optical microscopy of the conventional cast and stir cast hypereutectic alloy has shown that stir casting causes refinement of primary silicon particles and modification of eutectic silicon compared to conventional cast alloy [6]. The formation of TiB_2 particles occurred via diffusion of Boron atoms through TiAl3 particles interface, thereby reacting to form fine TiB2 particles [7]. The use of aluminium-based particulate reinforced MMCs for automotive components and aircraft structures have been shown to be highly advantageous over their unreinforced alloys, due to their high specific strength and stiffness and superior wear characteristics in a wide temperature range. SEM analysis of the fracture surfaces of the tensile specimens showed similar morphologies for the cast and forged composites. The mechanism of damage was mainly decohesion at the matrix–particle interface [8]. Rheological properties of aqueous concentrated AlN suspensions have been investigated in the presence of a sintering aid, deffloculant, binder and plasticizers, in order to screen the most suitable experimental conditions to obtain a good rheological behavior for tape casting thick and non-cracked tapes with good flexibility [9]. The wear rate of all heat-treated specimens is less than that of the specimens in the cast condition. The main reason for this improvement is clearly related to the hardness enhancement after the aging treatment[10]. The wear resistance of Al–Si alloy matrix composites has increased with increasing wt% of Al_2O_3 particles. This is due to higher load-carrying capacity of the hard reinforcement Al_2O_3 particles, which limits the amount of plastic deformation of the matrix [11]. At the initial stage of heating, any moisture in the ceramic particles and the matrix materials burns off, so it reduces the level of porosity. The mechanical properties like hardness, tensile strength of A359/Al_2O_3 MMC increased [12]. Al 7075/Al_2O_3/SiC MMC resulted high micro hardness and low amount of wear reported [13]. The density was reduced but hardness porosity increased with alumina percentage in Al356/Alumina/ fly ash hybrid MMC [14]. Stir casting set up yields uniform mixing of 6% wt. fly ash particles and Al MMC [15]. With the increase in SiCp particles in Al 7075 alloy the porosity in the composite increased. SiCp particles enhanced hardness in MMC [16]. Porosity was increased by an increase in % of reinforcement in AA6063/ fly ash MMC [17]. Literature survey did not reveal any systematic study of Al 6064/Al_2O_3 metal matrix composites. Objective of the present work Fabrication of Al6064/Al_2O_3 metal matrix composites with different percentage of reinforcement by stir casting method. Microstructure observation of all fabricated samples to observe distribution of reinforcement and Investigation of mechanical properties of all samples.

2. Experimentation

2.1Matrix material

Most of the automobile parts, aerospace structures and its allied infrastructure are made of aluminum alloy. Al 6064 alloy is used for making particulars for braking systems for automotive and electronics industry. Al 6064 has good machinability and anodizing characteristics. It has suitable electrical and thermal conductivity. The chemical composition of Al 6064 is

Table 1. Elemental composition of aluminum alloy (Al 6064)

Element	Composition % in Al 6064
Cu	0.4
Mg	1.2
Si	0.8
Fe	0.7
Mn	0.15
Al	Remaining

2.2 Reinforcement material

Alumina is widely used as the reinforcement in the metal matrix composites. The mechanical and physical properties of alumina (Al_2O_3) have been reviewed from literature data for the

IOP Conf. Series: Materials Science and Engineering **802** (2020) 012004 doi:10.1088/1757-899X/802/1/012004

purpose of characterizing the mechanical properties of alumina for manufacturing process in engineering application.

Table 2. Mechanical properties of (Al_2O_3)

Alumina	Grade	pH value	Mesh size
Neutral(1344-28-1)	Brockman 1	6.5-7.5	75μm

Metal–matrix composites (MMCs) are most promising in achieving enhanced mechanical properties such as: hardness, toughness and ultimate tensile strength due to the presence of micro-sized reinforcement particles into the matrix. Generally, the reinforcements result in higher strength and hardness, often at the expense of some ductility. MMCs reinforced with alumina are widely used for high performance applications such as in automotive, aerospace and electrical industries because of their improved physical and mechanical properties.

2.3 Experimental procedure

The materials used in the present work are aluminium alloy (Al 6064) and aluminium oxide (as reinforcement). These materials are chosen due to their easily mixable property and gives good mechanical properties. The aluminium alloy (Al 6064) is heated up to its melting temperature 600^0C in a muffle furnace. Temperature measurement is done with help of K-type thermocouple. Preheated aluminium oxide is mixed slowly in molten aluminium alloy with the help of stirrer. The material is left for cool down in the crucible in which it was melted and mixed. There are three samples in different ratios which are prepared for testing mechanical properties. The ratios are shown below in the Table 3.

Table 3. Composition selection

Material	Sample 1	Sample 2	Sample 3
Al 6064	97.5%	95%	92.5%
Alumina	2.5%	5%	7.5%

Figure 1 a. Muffle furnace **Figure 1 b.** Preparation of sample by stir casting

Now the mixed composite after cooling and solidification is cut into desired shapes on the band saw machine. The machine is an electrically driven machine which is used to cut the hard composites. The MMC samples prepared are shown in figures given below

Figure 2. MMC (Al 6064+2.5% of Al_2O_3)

Figure 3. MMC (Al 6064+ 5% of Al_2O_3)

Figure 4. MMC (Al 6064+7.5% of Al_2O_3)

Figure 5. MMC sample prepared for hardness test

Figure 6. MMC sample for tensile test

Figure7. MMC sample prepared for toughness test

Figure 8. MMC sample prepared for microstructure

The prepared samples had an uneven surface. To obtain the smooth surface for microstructural examination, the samples were then taken for grinding/polishing operation. The grinding was done in such a way so that all the scratches are in the same direction and the grinded surface becomes flat. After this the samples were polished using different grits of emery papers. Because the samples being aluminium alloy composites it is rubbed over the emery paper for a small time. It was rubbed over an emery paper of 400, 600, 800, 1000 grit and then over a very fine emery paper of 1200, 1600, 2000 grit for a considerable time in order to get a smooth and clear surface of the samples.

3. Results and Discussion

3.1 Analyses of microstructure

The microstructures of the MMC samples are seen using metallurgical microscope. The term 'microstructure' in metal matrix composites is used to describe the appearance of the reinforcement material. A reasonable working definition of microstructure is "The arrangement of phases and defects within a material." Samples of as cast MMCs for microstructural examination were prepared by grinding through different sizes of grit papers. Then the samples were etched with the etchant i.e., Keller's reagent (2.5mL nitric acid, 1.5mL HCl, 1.0mL HF, 95.0 ml water) [12]. The microstructure of etched samples was observed by using Metallurgical Microscope. It is seen that for all samples the alumina particles are evenly distributed in the aluminium matrix. Figure 9 (b) and (c) shows the accumulative nature of alumina particles as their amount increases.

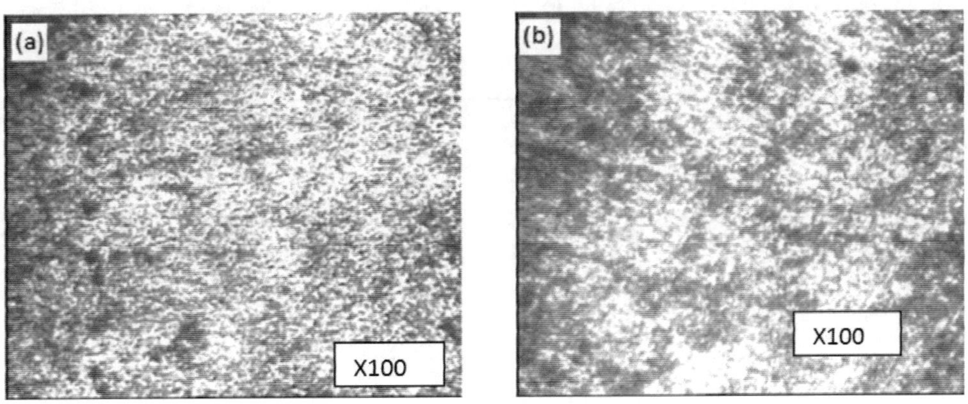

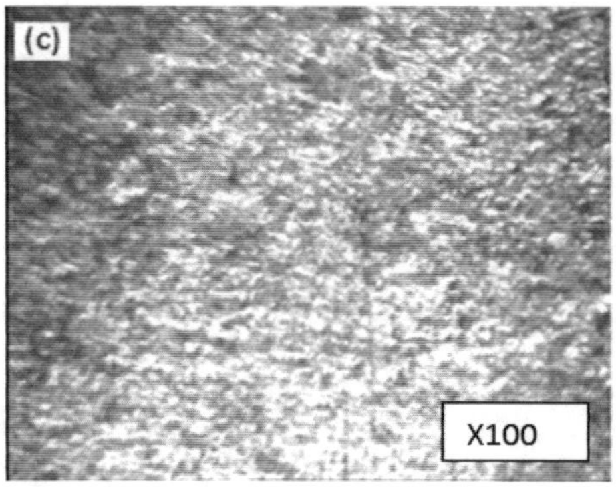

Figure 9. (a) Al6064+2.5 % Al_2O_3 reinforcement **(b)** Al6064+5 % Al_2O_3 reinforcement **(c)** Al6064+7.5 % Al_2O_3 reinforcement

3.2 Analysis of tensile strength

Tensile tests are used to determine the elastic limit, elongation, tensile strength, yield strength and reduction in area. Tensile tests were carried out at room temperature using computerized universal testing machine. The ultimate tensile strength, percentage elongation was calculated for each prepared sample. The specimens were tested according to ASTM E8M-01 standard.

Table 4. Tensile Test of Al 6064+Al_2O_3 MMC

Samples (of Al_2O_3)	Ultimate strength(N)	Elongation (mm)	Break load (N)	Break Elongation (mm)	True UTS (N/ sq mm)	Area (sq. mm)
2.5%	699.8	0.53	148.3	0.63	26	28.386
5.0%	637.5	0.5	205.9	0.67	22.9	28.186
7.5%	1398.1	2.16	1008.3	2.50	62.2	26.123

3.3 Analysis of toughness test

In these MMC specimens toughness is tested by breaking it with impact force of a hammer weighing 21 kg. The Hammer is leaved from 140 degree of angle with the initial energy of 300 J. According to ASTM E-23, standard specimen, sizes are 10 mm x 10 mm x 55 mm. for charpy impact test analysis. The table shows the results while testing as follows:

Table 5. Toughness of Al 6064+Al_2O_3 MMC

Samples (of Al_2O_3)	Trial			Average Energy (Joule)
	A	B	C	
2.5%	41	43	38	40.66
5.0%	37	33	35	35
7.5%	26	24	28	26

3.4 Analysis of hardness test

The hardness test was carried on Rockwell Hardness testing method. For hardness testing, the samples of Al 6064/2.5, 5, 7.5 % wt. Al_2O_3 composite were prepared as per dimension (10 mm x 10 mm x 25 mm) as ASTM E18-05, standard.

Figure 10. MMC sample after indentation on hardness testing machine

Table 6. Hardness of MMC

Samples (of Al_2O_3)	Trial A	Trial B	Trial C	Mean Hardness
0%	94	96	98	96
2.5%	105	102	108	108
5.0%	122	118	114	118
7.5%	137	142	145	143.33

4. Conclusion

The following conclusions are arrived based on the experimental investigation on the distribution of Al_2O_3 in the mechanical stir casting and its effect on mechanical properties of the as cast MMCs at different weight fraction of Al_2O_3.

1. In the present investigation, it was observed that maximum mechanical properties were obtained for Al 6064/7.5% Al_2O_3.
2. The tensile strength of the cast composites increases on increasing the weight fraction of Al_2O_3.
3. The hardness of the MMCs is higher than the unreinforced matrix metal and the hardness of the cast composites increases linearly with increasing the weight fraction of Al_2O_3.
4. Microstructural observation suggests that stirring action produces MMC with smaller grain size and there is a good particulate matrix interface bonding.
5. The increase in Al_2O_3 constituent toughness decreases with respect to base metal. This is due to increase in brittleness between the alloy Al 6064 and Al_2O_3 interfaces.

References

[1] Rajmohan T and Palanikumar T 2013 Application of the central composite design in optimization of machining parameters in drilling hybrid metal matrix composites *Measurement* **46**, 1470–1481

[2] Naher S, Brabazon D and Looney L 2007 Computational and experimental analysis of particulate distribution during Al–SiC MMC fabrication. *Composites: Part A* **38** 719–729

[3] Tzamtzis S, Barekar N S, Hari Babu N, Patel J, Dhindaw B K and Fan Z 2009 Processing of advanced Al/SiC particulate metal matrix composites under intensive shearing-A novel Rheo-process *Composites: Part A* **40** 144–151

[4] Chaudhury S K, Singh A K, Sivaramakrishnan C S and Panigrahi S C 2005 Wear and friction behavior of spray formed and stir cast Al–2Mg–11TiO2 composites *Wear* **258** 759–767

[5] Karamis M and Nair F 2008 Effects of reinforcement particle size in MMCs on extrusion die wear *Wear* **265** 1741–1750

[6] Reddy T V S, Dwivedi D K and Jain N K 2009 Adhesive wear of stir cast hypereutectic Al–Si– Mg alloy under reciprocating sliding conditions. *Wear* **266** 1–5

[7] Emamy M, Mahta M and Rasizadeh J 2006 Formation of TiB2 particles during dissolution of TiAl3 in Al–TiB2 metal matrix composite using an in situ technique *Composites Science and Technology* **66** 1063–1066

[8] Ceschini L, Minak G and Morri A 2009 Forging of the AA2618/20 vol.% Al_2O_3p composite:Effects on microstructure and tensile properties *Composites Science and Technology* **69** 1783–1789

[9] Olhero S M and Ferreira J F 2005 Rheological characterization of water-based AlN slurries for the tape casting process *Journal of Materials Processing Technology* **169** 206–213

[10] Erdem K, Muzaffer Z and Ridvan Y 2013 Effect of heat treatment conditions on microstructure and wear behaviour of Al4Cu2Ni2Mg alloy *Trans Nonferrous Met Soc China* **23**

[11] Kala H, Mer KKS and Kumar S 2014 A review on mechanical and tribological behaviors of stir cast aluminum matrix composites *Procedia Mater Sci* **6** 1951–1960

[12] Kumar Abhishek, Lal Shyam and Kumar Sudhir 2013 Fabrication and characterization of A359/Al_2O_3 metal matrix composite using electromagnetic stir casting method *Journal of Materials Research and Technology* **2** (3), 250-254

[13] Suryakumari T S A, Ranganathan S and Shankar P 2015 Study on mechanical properties of Al 7075 hybrid metal matrix Composites *Applied Mechanics and Materials* **230** 813–820

[14] Kulkarni S G, Menghani J V and Lal A. 2016 Investigation of mechanical properties of fly ash and Al_2O_3 reinforced A356 alloy matrix hybrid composite manufactured by improved stir casting *Indian Journal of Engg. &Material Science* **23** 27–36

[15] SharmaV K, Singh R C and Chaudhary R. 2017 Effect of flyash particles with aluminium melt on the wear of aluminium metal matrix composites *Engineering Science and Technology an International Journal* 18–23

[16] Das D, Chakraborty V, Nanda B K and Routara B C 2018 Turning performance of Al 7075/SiCp MMC and multi-response optimization using WPCA and Taguchi approach *Materials Today Proceedings* **5** 6030

[17] Razzaq M, Alaa DM, Ishak M and Basheer U 2017 Effect of fly ash addition on the physical and mechanical properties of AA6063 alloy reinforcement *Metals* **7** 477

Effect of TiO2 additive volume fraction in lubricant oil on the performance of hydrodynamic journal bearing

Anurag Singh, Neeraj Verma*, Aman Chaurasia, Alok Kumar,
Department of Mechanical Engineering, JSS Academy of Technical Education, Noida, India
*E-mail: neerajverma@jssaten.ac.in

Abstract
Hydrodynamic journal bearings are widely used in rotating machines where high loading condition prevails. In this work, effect of different volume fractions of TiO2 (Titanium dioxide) additive in commercially available SAE30 lubricant oil on the performance of hydrodynamic journal bearing is analysed experimentally. Hydrodynamic journal bearing test rig is used to perform the experiment. The experiment is carried out by taking different volume fractions of TiO2 additive (0.075%, 0.1% and 0.15%) in SAE30 lubricant base oil and their effect on journal bearing characteristics like coefficient of friction, pressure distribution, temperature rise, eccentricity ratio, attitude angle, minimum film thickness and displacement is analysed. Paraffin oil is used to uniformly distribute the TiO2 in the base oil and to resist sedimentation.
The study shows that the addition of TiO2 additive in the base SAE30 lubricant oil is capable of significantly reducing the values of friction coefficient and maximum temperature rise. It has been found that the addition of TiO2 additive improves the performance of the hydrodynamic journal bearing. Results reveals that the additive performs better at higher loading conditions.

Key words: Hydrodynamic journal bearing, Coefficient of friction, Pressure, Load, Temperature

1.Introduction

Hydrodynamic journal bearings are used in compressors, pumps, turbines, internal combustion engines, motors, generators etc. due to their high load carrying capacity and good damping properties [1-3]. The main function of journal bearing is to support radial load and facilitate motion as well as transfer of power [4][5].
Additives are introduced in the base stock to improve their tribological performance, thermal properties and anti-oxidation capability [6][7]. Like other additives TiO2 additives are also used to improve the performance of base oil or lubricant. It is capable of enhancing the load carrying capacity [8] and reducing the friction and wear [9-11]. Tribofilm formation on the sliding surface and ball bearing effect of the TiO2 particles are primary reason behind it [12]. Surfactants and dispersants are generally used to disperse additives in the base oil uniformly and to avoid their sedimentation [13]. Thus, in this study paraffin oil is used along with the pure oil to uniformly distribute the TiO2 additive.

2. Experimental study

2.1 Sample preparation

TiO2 (Titanium Dioxide) additives used in this study is purchased from Qualikems Fine Chem Pvt. Ltd. (Nandesari, Vadodara, India). SAE30 oil as exhibited in figure 1(a) is used as a base oil in the experiment and TiO2 additive is blended in it. Figure 1(b) shows the SAE30 oil mixed with TiO2 additive.

(a) (b) (c)
Fig. 1 (a) Pure SAE30 lubricant oil, (b) SAE30 oil mixed with TiO2 additive and (c) Paraffin oil

To hinder the sedimentation and aggregation of the TiO2 particles and for its homogenous mixing in the base oil, Paraffin oil as shown in figure 1(c) is used as dispersant [14]. Table 1 shows the properties of SAE30 lubricant oil.

Table 1 Properties of lubricant SAE30

Property	Value in metric unit	
Density at 20°C	$0.885*10^3$	Kg/m^3
Kinematic Viscosity at 40°C	85.76	cSt
Kinematic Viscosity at 90°C	11.42	cSt
Viscosity Index	100	-
Flash Point	210	°C
Pour Point	-23	°C
Sulfated Ash	1.67	%

The experiment is performed by taking pure SAE30 oil and then subsequently adding different volume fractions (0.075%, 0.1% and 0.15%) of the TiO2 additive in the base oil. The calculation for the mass of TiO2 has been carried out by using equation 1 [15].

$$m_{TiO_2} = \rho_{TiO_2} \times \phi \times V_l \qquad (1)$$

Where, m_{TiO_2} is the mass of TiO2 additive (g)

ρ_{TiO_2} is density of additive TiO_2 (kg/m^3)

ϕ is volume fraction

V_l is Volume of lubricant in litre

The calculated amount of TiO2 is weighed on weighing machine and is then mixed with paraffin oil dispersant. Paraffin oil is 15% by volume of base lubricant oil. The mixture is allowed to mix using magnetic stirrer for 30 minutes. Then it is dispersed in the SAE30 base lubricant oil and is mixed for two hours.

2.2 Hydrodynamic journal bearing test rig

Journal bearing test rig shown in figure 2 consists of a journal having diameter 100 mm, mounted horizontally on a housing supported on self-aligned bearing. The one side of journal is connected to 3 phase induction motor with toothed belt drive, on the other side the test bearing is fitted. The other end of journal/shaft is mounted above the frame to rotate in the horizontal axis.

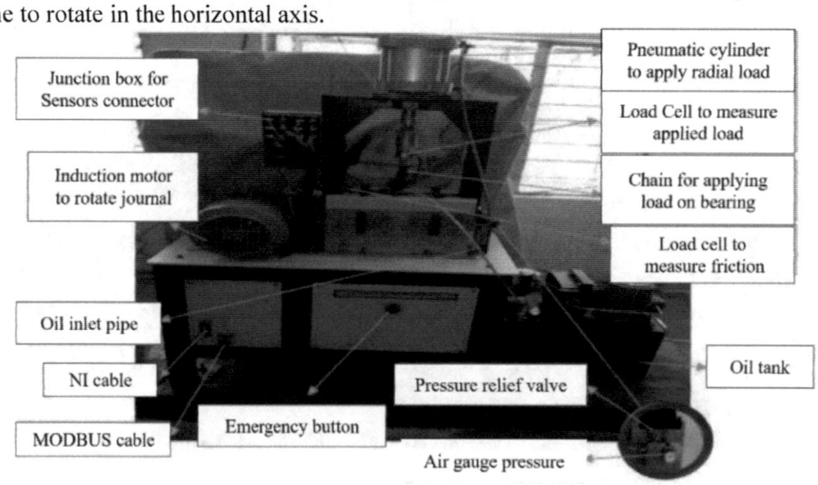

Fig. 2 Hydrodynamic Journal Bearing Test Rig (HJBTR)

IOP Conf. Series: Materials Science and Engineering **802** (2020) 012005 doi:10.1088/1757-899X/802/1/012005

A pneumatic cylinder is fitted above the frame to apply load on bearing. Air pressure is generated by air compressor. A reciprocating double acting piston is fitted in the cylinder and connected to the bearing with load cell and chain. Difference of air pressure in upper and lower part of piston created pressure difference across the piston. This pressure difference causes load on the piston which is transformed to the bearing via connected load cell and chain fitted to the bracket on the bearing. Load cell calculate the load, displays on software screen. Pneumatic cylinder is capable to sustain maximum load of 10kN when pressure in the pneumatic pressure gauge shows 6 bar.

An oil tank of capacity 20 litres is used to store the lubricant. A 3-phase induction motor is used to pump the lubricant. Pumped lubricant is cooled by a cooling fan arrangement placed over oil tank. Pressure of pumped lubricant can be regulated by pressure regulator valve. Pumped lubricant reach to journal bearing arrangement via inlet pipe, circulate within it. Side flow of lubricant come back to the oil tank via drain pipe, filtered at inlet of oil tank. The lubricant stored in oil tank can be circulate again and again (figure 3).

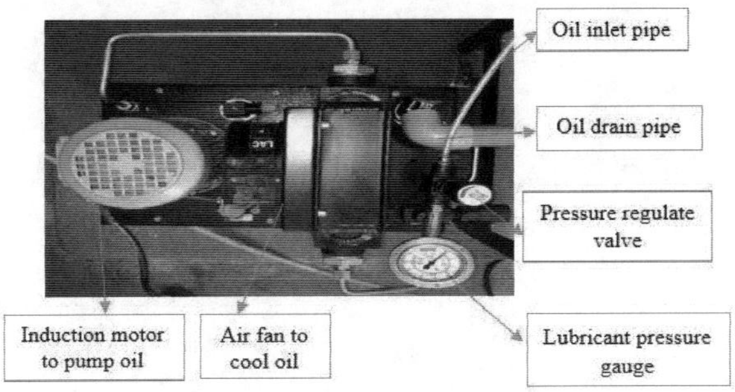

Fig. 3 Lubricant supply system

Fig. 4 Sensors installed in HJBTR

6 pressure sensors and 6 temperature sensors mounted on the bearing as illustrated in figure 4 are used to measure pressure and temperature. Lubricant reach between journal-bearing arrangements via entry hole. Also, entry holes are made on bearing to reach the lubricant at various sensors. Variable frequency drive is used to control the speed of 3-phase induction motor. WINDUCOM 2010 software is used to control the journal bearing test rig. An NI cable and a MODBUS cable is used to interface between computer and journal bearing test rig. MODBUS cable send command from computer to different components of machine and the components responds accordingly. NI cable send the response of different sensors of machine to the computer software. Table 2 shows the specifications and geometric properties of HJBTR.

Table 2 Specifications and geometric properties of HJBTR

Speed Range	150 to 1500 RPM
Normal test load Range	Upto 10 kN
Maximum Pressure Range	68.9 bar
Temperature Range	Room Temperature to 200°C
Maximum Film Thickness	100 μm
Bearing Material	Brass
Shaft Diameter (d_S)	100.0 mm
Bearing Inner Diameter (d)	100.2 mm
Bearing Length (L)	100.0 mm
Radial Clearance (c)	0.100 mm

3. Result and discussion

The experiment is performed at a speed of 1400 rpm and the load varies from 600 N to 1000 N. The operating temperature inside the bearing is below 100^0 C and the pressure range inside the bearing is 0 to 7 bars.

3.1 Variation of coefficient of friction

Friction coefficient value is decreasing with increase in the volume fraction of the additives as exhibited in the figure 5. The coefficient of friction is minimum for 0.15% volume of TiO2 added in the SAE30 oil at all loading conditions. However, 0.1% volume fraction of the TiO2 is not following the trend.

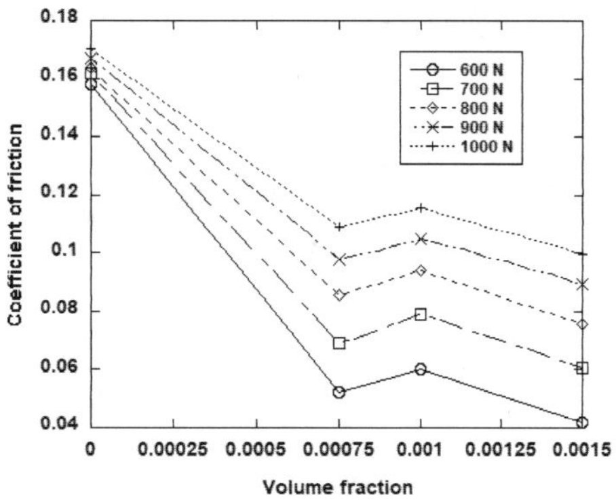

Fig. 5 Variation of coefficient of friction with volume fraction for different loads.

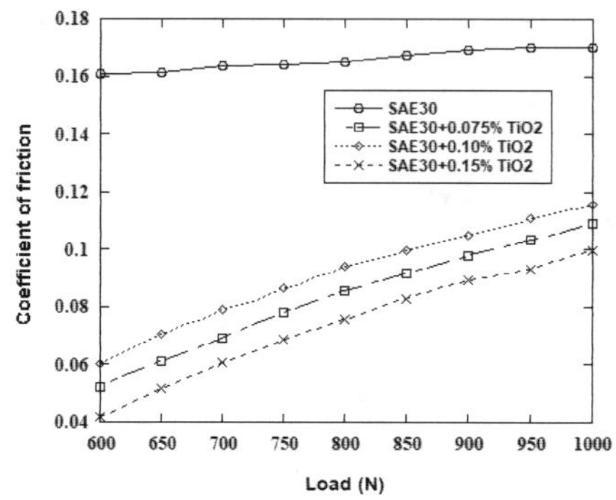

Fig. 6 Variation of coefficient of friction with load for different volume fractions of the TiO2 additive.

Figure 6 shows the variation of the friction coefficient with load. It can be seen from the graph that the friction coefficient is increasing with increase in load.
It's evident from the both figures (Figure 7 and Figure 8) that the COF value is decreased by 66.89%, 61.69% and 73.54% for 0.075%, 0.1% and 0.15% volume fraction of the TiO2 dissolved in the base SAE30 oil respectively at 600 N as compared to the pure base oil. The COF value is decreased by 37.25%, 32.05% and 41.234% for 0.075%, 0.1% and 0.15% volume fraction of the TiO2 dissolved in the base SAE30 oil respectively at 1000 N compared to the base oil. TiO2 particles form a tribofilm between the journal bearing surface, this tribofilm formation [29][30] prevents the direct metal to metal contact and thus reduces the friction values drastically.

3.2 Pressure distribution at different load for various volume fractions

Figure 7 (a-e) shows the variation of the Pressure with circumferential angle at different loads. Readings of pressure are taken from the six pressure sensors installed at a gap of 36^0 each. Each graph is plotted at constant load and for different volume fractions (0.075%. 0.1% and 0.15%) of TiO2 additive.
Maximum pressure or peak pressure is obtained at a circumferential angle of 108^0 for all loading conditions and each composition of the TiO2 additives added.
The peak pressure is maximum for SAE30 base lubricant oil than the peak pressure of SAE30 oil dissolved with different volume fractions of the TiO2 additives at 600N. However, with increase in the load the value of peak pressure for the SAE30 lubricant oil dissolved with TiO2 additives comes closer to the pure SAE30 oil. At initial loading condition, the value of peak pressure is decreasing with an increase in the TiO2 concentration in the base oil (figure 7(a)). Whereas, the gap in peak pressure is decreasing with increase in load. The difference in peak pressure between pure oil and base oil with 0.15% volume fraction of TiO2 is 0.004581 (N/m^2) at 600N. However, at 1000N the difference reduces to 0.002083 (N/m^2). Apparently, the TiO2 additives in base SAE30 oil show better results at higher loads.
The value of the peak pressure developed is improving as the load is increased for all compositions of the TiO2 additives in base oil and is also improving for the pure SAE30 oil. The development of pressure is conducive in keeping the journal bearing away from the contact of surface asperities.

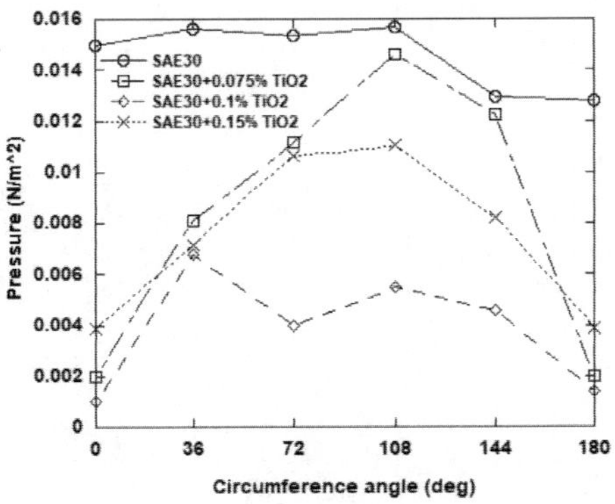

(a) 600N

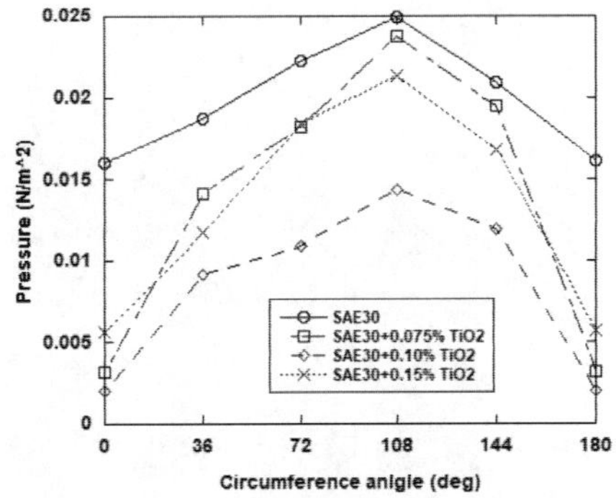

(b) 700N

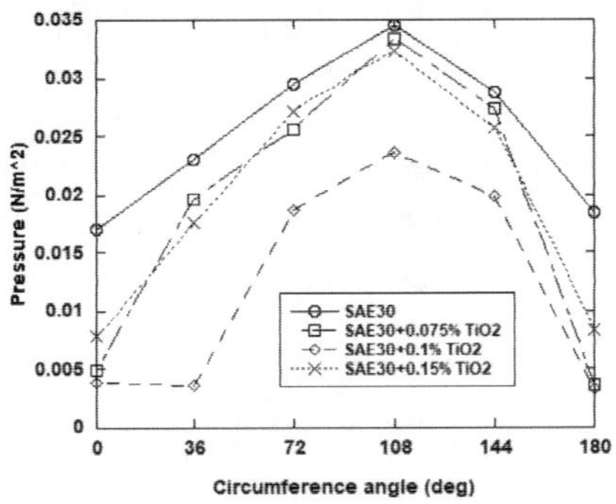

(a) 800 N

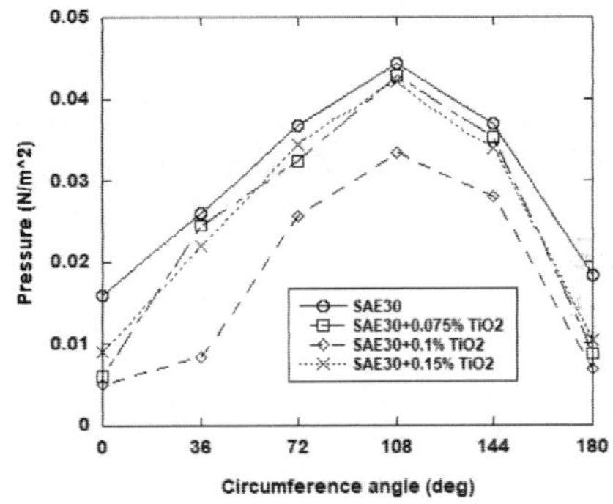

(b) 900 N

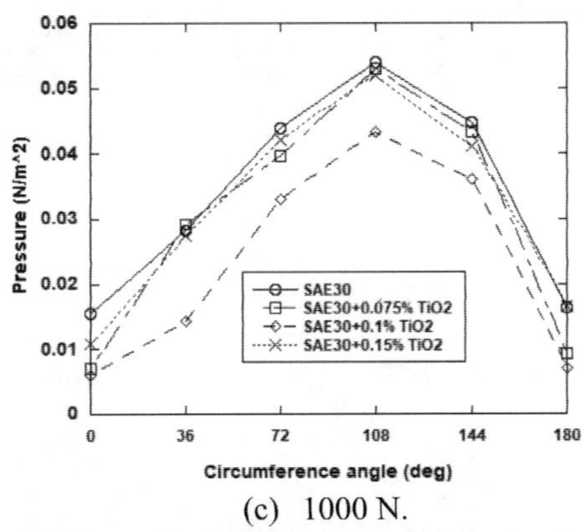

(c) 1000 N.

Fig. 7 Variation of Pressure with Circumference Angle for different volume fractions of the TiO2 added at (a) 600 N, (b) 700 N, (c) 800 N, (d) 900 N, and (e) 1000 N

3.3 Temperature distribution at different volume fraction.

Figure 8 is a bar graph showing temperature on the y-axis and the numbers on the x-axis from 1 to 5 are the temperature sensors installed at a gap of 36° each. Temperature sensor 3 installed at an angle of 108° is showing the maximum temperature than all other sensors. Indicating that the temperature is maximum at a circumferential angle of 108°.

Addition of the TiO2 particles in pure SAE30 oil reduces the maximum temperature value and its lowest for 0.1% volume fraction of TiO2 (Figure 9). The result indicates that addition of the TiO2 additive improves the heat dissipation rate.

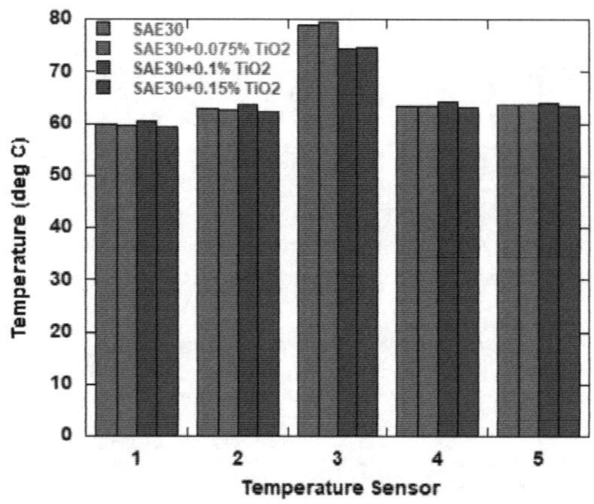

Fig. 8 variation of temperature for different volume fraction measured by sensors.

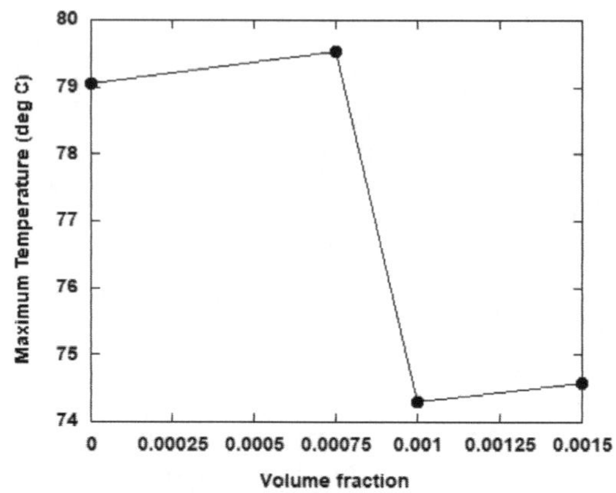

Fig. 9 Variation of maximum temperature rise vs volume fraction

3.4 Variation of other characteristics with volume fraction

Eccentricity ratio is initially increasing with increase in the volume fraction of the TiO2 additive and then further decreasing. It is minimum for pure base oil and maximum for 0.1% TiO2 added (Figure 10). Similar behavior is followed by the attitude angle vs volume fraction graph as depicted in figure 11 where attitude angle is initially increasing with addition of the additive and is then decreasing. For 0.15% volume fraction, the value of attitude angle is more compared to that of the 0.1% volume fraction.

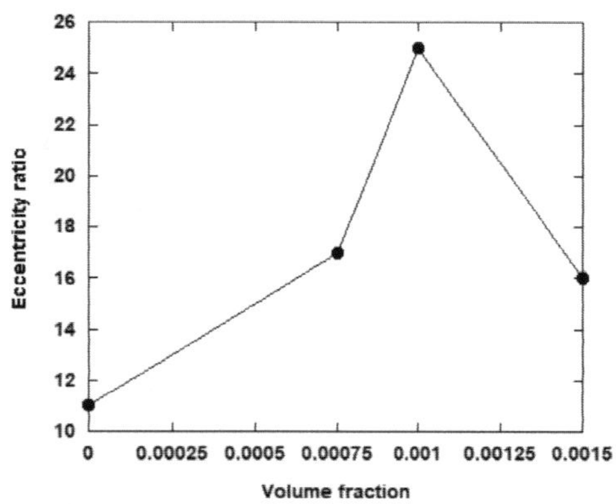

Fig. 10 Eccentricity ration vs volume fraction

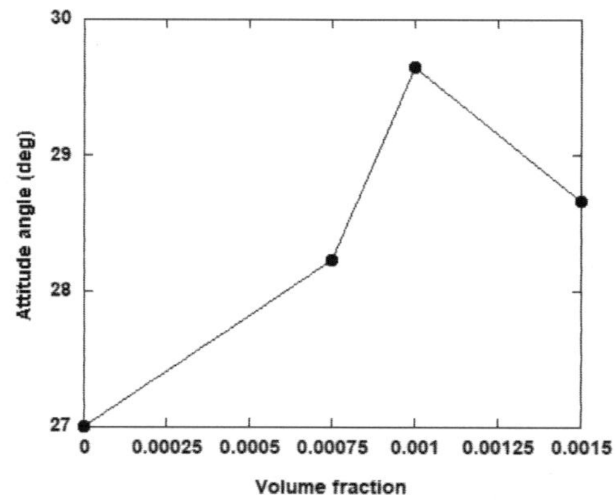

Fig. 11 Attitude angle vs volume fraction

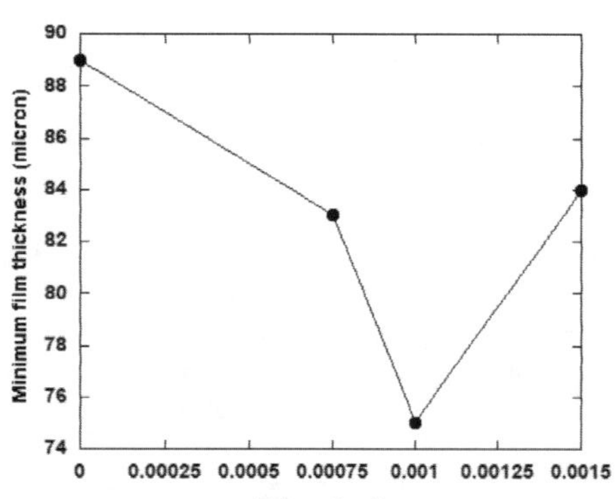

Fig. 12 Minimum film thickness vs volume fraction

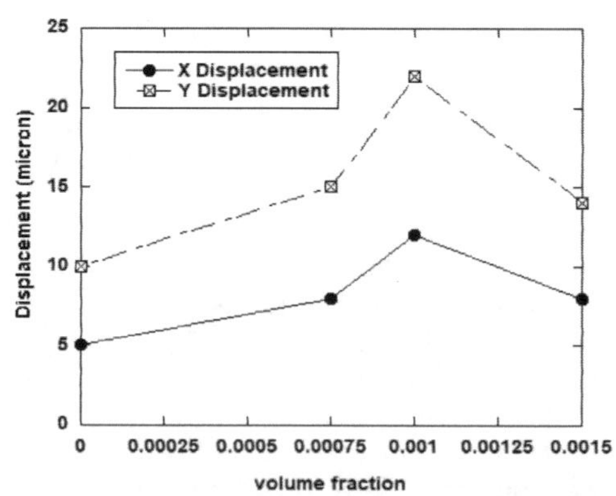

Fig. 13 Displacement vs volume fraction

Minimum film thickness decreases with increase in the volume fraction of the additive up to 0.1% TiO2 and is then increasing (Figure 12). In a journal bearing operation, minimum fluid film thickness plays a crucial role to deter the direct metal to metal contact [16]. In this case, film thickness is maximum for pure SAE30 base lubricant oil and minimum for 0.1% TiO2 additive. X and Y displacement of journal bearing is illustrated in figure 13. Both X and Y displacement are maximum for 0.1% volume fraction of TiO2 additive and is minimum for pure SAE30 lubricant oil.

4.Conclusion

Addition of the TiO2 additive in the SAE30 base oil can significantly reduce the value of the friction coefficient. The value of the peak pressure developed is improved as the load is increased for all compositions of the TiO2 additives in base SAE30 oil. The result shows that TiO2 additives perform better at higher load. Addition of the TiO2 particles in pure SAE30 oil reduces the maximum temperature raised in the journal bearing. However, value of minimum thickness is maximum for pure oil. Eccentricity ratio, attitude angle and displacement values are minimum for SAE30 base oil and is maximum for 0.1% volume fraction of TiO2 additive. The characteristics displayed by 0.1% TiO2 additive volume fraction highly discourages its use in journal bearing. Whereas, 0.15% TiO2 is displaying more amicable results compared to others.

Acknowledgement

The authors appreciate the support of Dr. Suresh Verma (Professor, Mechanical Engineering Department, DCRUST Murthal) and Priyesh Ranjan (M.Tech. DCRUST Murthal).

References

1. Kalani, A., Soni, S., & Jani, R. (2015). Expert Knowledgebase System for Computer Aided Design of Full Hydrodynamic Journal Bearing. *International Journal of Mechanical Engineering and Technology*, *6*(8), 46-58.
2. Kerlekar, R. S., Dhande, D. Y., Nigade, M. B., & Kondhalkar, G. E. (2015). Performance Analysis of 3 Lobe Hydrodynamic Journal Bearing. *International Journal of Engineering Research and General Science*, *3*(4).
3. Sharma, S., & Awasthi, R. K. (2017). Effect of aspect ratio on the performance of hydrodynamic journal bearing operating under wear. *International Journal of Theoretical and Applied Mechanics*, *12*(3), 497-522.
4. Kasolang, S., Ahmad, M. A., Joyce, R. D., & Tai, C. F. M. (2012). Preliminary study of pressure profile in hydrodynamic lubrication journal bearing. *Procedia Engineering*, *41*, 1743-1749.
5. Shahnazar, S., Bagheri, S., & Hamid, S. B. A. (2016). Enhancing lubricant properties by nanoparticle additives. *International journal of hydrogen energy*, *41*(4), 3153-3170.
6. Krishna, P. V., Srikant, R. R., Padmini, R., & Viswaditya, J. L. P. P. (2013, July). Application of nanomaterials as coolants/lubricants in machining. In *International Conference on Advanced Nanomaterials & Emerging Engineering Technologies* (pp. 674-682). IEEE.
7. Singh, A., Chauhan, P., & Mamatha, T. G. (2019). A review on tribological performance of lubricants with nanoparticles additives. *Materials Today: Proceedings*.
8. Binu, K. G., Shenoy, B. S., Rao, D. S., & Pai, R. (2014). A variable viscosity approach for the evaluation of load carrying capacity of oil lubricated journal bearing with TiO2 nanoparticles as lubricant additives. *Procedia materials science*, *6*, 1051-1067.
9. Ilie, F., & Covaliu, C. (2016). Tribological properties of the lubricant containing titanium dioxide nanoparticles as an additive. *Lubricants*, *4*(2), 12.
10. Ingole, S., Charanpahari, A., Kakade, A., Umare, S. S., Bhatt, D. V., & Menghani, J. (2013). Tribological behavior of nano TiO2 as an additive in base oil. *Wear*, *301*(1-2), 776-785.
11. Xia, W., Zhao, J., Wu, H., Jiao, S., Zhao, X., Zhang, X., ... & Jiang, Z. (2018). Analysis of oil-in-water based nanolubricants with varying mass fractions of oil and TiO2 nanoparticles. *Wear*, *396*, 162-171.
12. Wu, H., Zhao, J., Cheng, X., Xia, W., He, A., Yun, J. H., ... & Jiang, Z. (2018). Friction and wear characteristics of TiO2 nano-additive water-based lubricant on ferritic stainless steel. Tribology International, 117, 24-38.
13. Padgurskas, J., Rukuiza, R., Prosyčevas, I., & Kreivaitis, R. (2013). Tribological properties of lubricant additives of Fe, Cu and Co nanoparticles. Tribology International, 60, 224-232.
14. Wu, Y. Y., & Kao, M. J. (2011). Using TiO2 nanofluid additive for engine lubrication oil. Industrial Lubrication and Tribology, 63(6), 440-445.
15. Sepyani, K., Afrand, M., & Esfe, M. H. (2017). An experimental evaluation of the effect of ZnO nanoparticles on the rheological behavior of engine oil. *Journal of Molecular Liquids*, *236*, 198-204.
16. Verma, S., Kumar, V., & Gupta, K. D. (2016). Effect of elasticity on capillary compensated flexible multi-recess hydrostatic journal bearing operating with micropolar lubricant. *Journal of The Institution of Engineers (India): Series C*, *97*(1), 11-23.

Simulation of CoFRP analytical and computational model for various fiber volume fraction and Investigation of the local stresses for different fiber orientation angle

Peeyush Maheshwari[1*], Anadi Misra[1], Ashtosh Kumar Singh[1], Kamal Singh Rawat[2], Ankur Verma[2] and Kavita Singhal[3]

[1]Department of Mechanical Engineering, G.B.P.U.A.T., Pantnagar, UK, India-263145.
[2]Department of Mechanical Engineering, M.I.E.T., Meerut, UP, India-250005.
[3]Department of Chemistry, G.B.P.U.A.T., Pantnagar, UK, India, India-263145.

*Email: peeyush.maheshwari@outlook.com

Abstract. Present work provides a numerical tool for analyzing the variation of failure strength of Coir Fiber Reinforced Polymer (CoFRP) with respect to Fiber Volume Fraction (FVF) and Fiber Orientation Angle (FOA). A model has been developed for analyzing the failure strength of CoFRP. It is analyzed for various fiber orientation angle and fiber volume fraction. Failure strength of the CoFRP depends upon both FVF and FOA. In this analysis, it has been investigated that the analytical results for CoFRP lamina are very close to computational results. Finite element analysis scores over the experimental tests as it is more convenient and economical method of analysis. Therefore, this method is more suitable for future work.

1. Introduction

Nowadays, Natural Fiber Reinforced Polymers are extensively used in structures in engineering and industrial applications which are subjected to loads. Natural Fiber Reinforced Polymer consists of two phases i.e. natural fiber and epoxy. In this investigation process, coir fiber is used as a reinforcement phase and epoxy as a protection and binding phase. Coir fiber and epoxy phase are not soluble in each other. All the CoFRP have behavior like orthotropic materials and have nine independent elastic constants i.e. three modulus of elasticity, three modulus of rigidity and three Poisson's ratio. Orthotropic materials have the maximum and minimum load carrying capacity in the longitudinal and transverse direction respectively [1]. The variation of failure strength has been determined through coding in both, MATLAB and ANSYS.

Mathematical equations and their curve are solved by generating a code with the help of MATLAB numerical tool. In current analysis micro-mechanics of lamina is used for computing the failure strength of CoFRP.

Different types of modelling and thermal analysis on the basis of their dimensions and boundary conditions are done by ANSYS which works on finite elements method [3].

The macro-mechanical analysis of fiber oriented lamina, independent from separate properties of fiber and matrix, is performed by considering lamina as a homogeneous and orthotropic composite material [4, 5].

IOP Conf. Series: Materials Science and Engineering **802** (2020) 012006 doi:10.1088/1757-899X/802/1/012006

2. Material and Model

The elastic constants and failure strength of Coir Fiber and Epoxy are shown in Table1.

Table 1. Material properties of natural fiber and epoxy

	Coir	**Epoxy**
E(GPa)	5.00	3.45
Nu	0.30	0.35
Sy(MPa)		16.00
Su(MPa)	175.00	24.00
ET(GPa)		0.28
G(GPa)	1.92	1.28

In the present simulation a model has been developed for the CoFRP. For the analysis of failure strength, the fiber volume fraction and the fiber orientation of lamina are taken from 30% to 70% and from 0° to 90° respectively. Lamina is fixed at the center. All degrees of freedom are zero at the center. Pressure is applied at the edge of the Natural Fiber Reinforced Polymer.

3. Previous Works

Nirbhay *et al.* (2014) determined the failure load for laminated composite under tensile loading. In this simulation the analysis of failure stresses of laminated composite has been carried out using ABAQUS. [6]

Rahmani *et al.* (2015) developed a mathematical technique to investigate the elastic constants of carbon fiber reinforced polymer laminates. For this purpose, a mathematical code was developed in MATLAB i.e. MATLAB calculation code (MCC). This code provides sufficient compatibility to analyze mechanical behavior of the composite material. The code in MATLAB uses various orientation angles of the fiber which makes it possible to analyze the volume fraction for a wide variety of fibers under mechanical load [7].

Maheshwari *et al.* (2018) also predicted the variation of elastic constants and stress of JFRP with fiber volume fraction and fiber orientation angle [8].

4. Methods

4.1. Macromechanical analysis for a lamina

A unidirectional lamina can be analyzed as an orthotropic material. It has nine independent constants.

4.1.1 Hook's law for 2-D lamina

A local co-ordinate system is used for generating stress-strain relationship. Two different axes of lamina, one along the direction of fibers and second perpendicular to the fibers are known as local longitudinal and local transverse axes respectively. In Figure1, global axes and local axes or material axes are shown by X-Y and 1-2 respectively [3].

4.1.2 Simulation of CoFRP lamina

Many FEM software are available for analysis such as ANSYS, ABAQUS, NASTRAN, FORTRAN etc. The simulation of CoFRP lamina has been done through ANSYS finite element analysis [5,8].

Finite element analysis requires these basic steps: pre-processing, processing and post-processing. In the first step the elastic properties, strength properties, lamina properties, geometry of lamina, meshing, boundary and loading conditions are defined. In processing step, ANSYS analyzes the conditions and solves the problem whereas results in terms of stress-strain distribution and failure strength or load are rendered in the post-processing step [2].

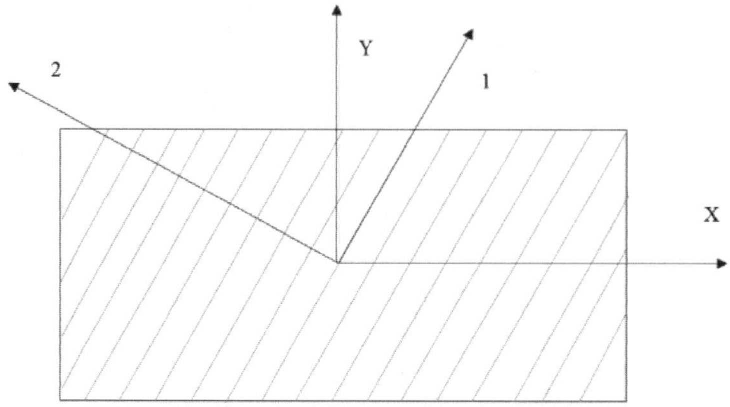

Figure1. Local and global axes

5. Results and Discussion

In the current macro-mechanical analysis of lamina, failure strengths of CoFRP are determined through two ways. Firstly, writing code in MATLAB and secondly computed using ANSYS software, then compared both results.

Figure 2 shows the variation of stresses in X, Y-direction (tensile) and XY plane (shear) with fiber volume fraction. The stress in longitudinal direction has higher value as compared to that in the transverse direction of lamina. Both, stress and shear strength, have linear variation with the fiber volume fraction and increase in stresses and shear strength is detected with the increment in fiber volume fraction.

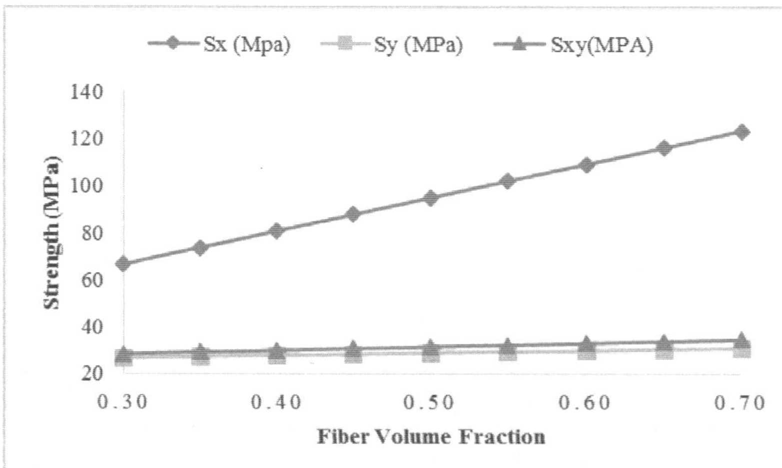

Figure2. Plot of Stresses vs FVF

5.1 Macro-mechanical Analysis of CoFRP

When orientation of fibers and loading direction was different, then CoFRP strength known as off-axis strength has been calculated.

Computational method, analytical method, theories of failure based on ANSYS, stress-strain relationship and MATLAB respectively have been used for calculated off-axis strength value of lamina and after that both the results are compared with each other.

5.2 Analysis of CoFRP at 50% fiber volume fraction

MATLAB program and ANSYS analysis has been used to calculate the failure strength and local stress value of lamina [8].

5.2.1 Comparison of computational (ANSYS) results and analytical (MATLAB) results

The comparison between failure strength values for lamina obtained from computational and analytical analysis presented in Figure 3 and are found to be in very good agreement with each other.

Figure3 shows the best optimum failure strength comparison results which obtained from lamina analysis by ANSYS.

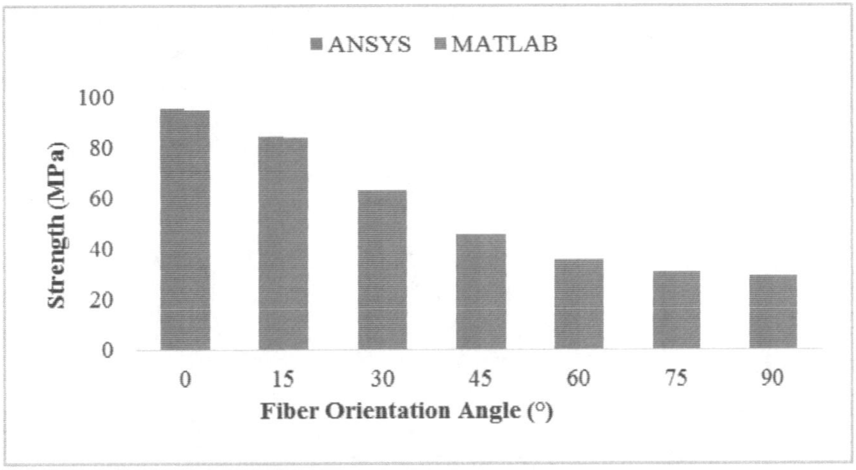

Figure3. Comparison of computational (ANSYS) and analytical (MATLAB) results

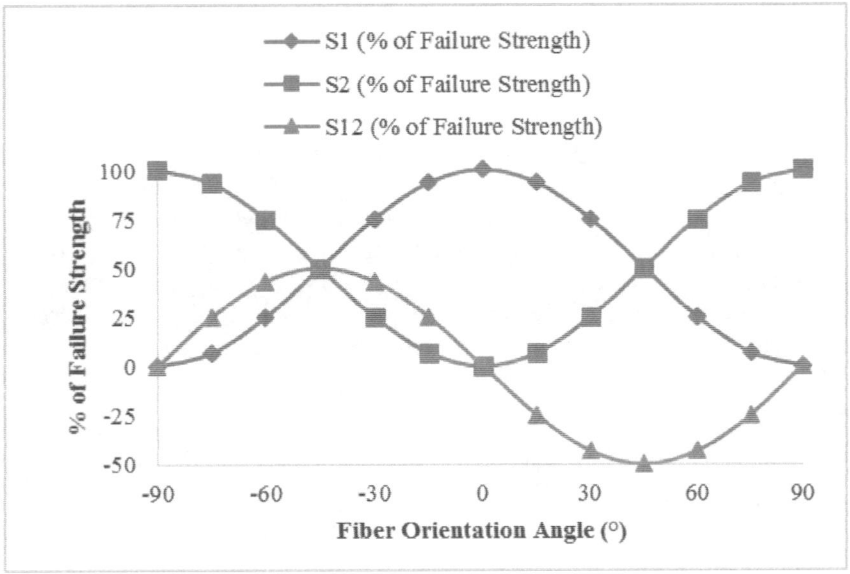

Figure 4. Plot of local stress vs FOA

IOP Conf. Series: Materials Science and Engineering **802** (2020) 012006 doi:10.1088/1757-899X/802/1/012006

Figure4 shows local stress variation with FOA. Increment of FOA in positive or negative direction decreased the value of failure strength in 1-direction. At ± 90° stress has been shown minimum value i.e. zero. Where as in 2-direction, Increment of FOA increased failure strength. It has maximum value at ±90.Shear failure strength in plane12 has minimum value at 0° and ±90°. The value of shear failure strength in plane12 increases up to -45° and after that decreases. Shear failure strength in plane12 has maximum value at -45°.

Figure5 shows local strain variation with FOA. Increment of FOA in positive or negative direction decreased the value of strain in 1-direction. At ± 90° strain has been shown minimum value i.e. zero. Where as in 2-direction, Increment of FOA increased strain. It has maximum value at ±90°. Strain in plane12 has minimum value at 0° and ±90°. The value of strain in plane12 increases up to -45° and after that decreases. Strain in plane1-2 has maximum value at -45°.

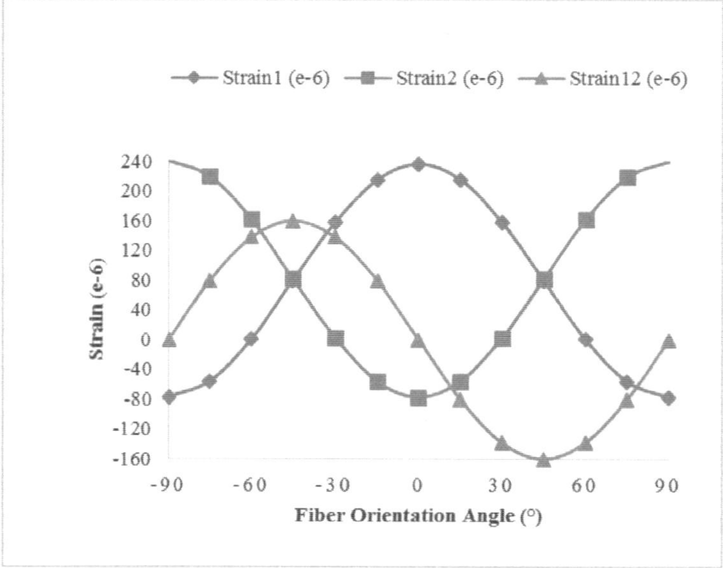

Figure 5. Plot of local strain vs FOA

6. Conclusion

In this study, the variation of stresses with fiber volume fraction has been determined. CoFRP has maximum longitudinal failure strength and lesser transverse failure strength. The determination of CoFRP is simulated using finite element analysis software ANSYS. Shell model was developed and analysis of the model was carried out as per given boundary conditions. Shell model was examined and the obtained results through ANSYS were in good agreement with the analytic results. Local stress and strain variation has been analyzed for CoFRP. Therefore analysis of the composite structure can be conveniently done with finite element method.

References

[1] Barbero E J 2007 *Finite Element Analysis of Composite Materials*, CRC Press, Boca Raton, 225p.

[2] Maheshwari P and Misra A 2017 Failure Test Simulation of Graphite Fiber Reinforced Polymer using Finite Element Analysis *International Journal of Innovative Trends in Engineering* **31** 29-33.

[3] Daniel I M and Ishai O 2006 *Engineering Mechanics of Composite Materials*, Oxford University Press, New York, 463p.

[4] Gowda T M, Naidu A C B and Chhaya R 1999 Some Mechanical Properties of Untreated Jute Fabric-Reinforced Polyester Composites *Composites: Part A* **30** 277-284.
[5] Reddy Y S N and Reddy J N 1992 Linear and non-linear failure analysis of composite laminates with transverse shear *Composite Science and Technology* **44** 227–55.
[6] Nirbhay M, Dixit A, Misra R K and Mali H S 2014 Tensile test simulation of CFRP test specimen using finite elements *International Conference of Advances in Manufacturing and Materials Engineering, AMME2014* **5(2014)** 267-273.
[7] Rahmani H, Najafi S H M and Ashori A 2015 Elastic Properties of Carbon Fibre-Reinforced Epoxy Composites *Polymers and Polymer Composites* **23** 475-481.
[8] Maheshwari P, Misra A and Kumar A 2018 Investigate the Variation in Elastic Constants and Failure Strength of Lamina with FVF and FOA for JFRP Lamina *International Journal of Applied Engineering Research.* **13** 157-161.

Effect of surface treatment on tribological characteristic of ferrite nanoparticles epoxy composites

Ashtosh Kumar Singh*[1], M. G. H. Zaidi [2,5], Rakesh Saxena[1], Rajeev Suman[3], Peeyush Maheshwari[1], H. P Gangwar[1], K. S. Rawat[4] and A. Verma[4]

[1]Mechanical Engineering Department, G.B. Pant University of Agriculture and Technology, Pantnagar, Uttarakhand, India
[2]Department of Chemistry, G.B. Pant University of Agriculture and Technology, Pantnagar, Uttarakhand, India
[3]Industrial and Production Engineering Department, G.B. Pant University of Agriculture and Technology, Pantnagar, Uttarakhand, India
[4]Mechanical Engineering Department, Meerut Institute of Engineering and Technology, Meerut, Uttar Pradesh, India

[5]Email: mgh_zaidi@yahoo.com

Abstract. In this investigation, surface treatment of ferrite nanoparticles were efficiently prepared by using amino propyl silane (APS). XRD which is a spectral analysis technique used for investigate the morphology and chemical structure of treated and normal ferrite nanoparticles. Ferrite nanoparticles epoxy composites were prepared by Modified ferrite nanoparticles (MFNs) with varying weight percentage and ferrite nanoparticles (FNs). XRD shown that MFNs was effectively modified with highly crystalline structure and silane surface treatment decreased the average particle size of FNs to ~16.48 nm. Surface treatment of FNs enhanced the homogeneous dispersion of ferrite in epoxy resin and improved interfacial adhesiveness between MFNs and epoxy. Tribological property of Modified ferrite nanoparticles epoxy composite (MFNEs) compare to cured epoxy and normal ferrite nanoparticles epoxy composites (FNEs) has been enhanced due to surface treatment.

Keywords. Ferrite Nanoparticles, Nano-magnetic Polymer Composites, Spectra, Tribological Characterization, Silane Coupling

1. Introduction

There is an expanding interest for multi-functional composites to encounter extraordinary prerequisites of engineering segments. [1–2]. Engineers and scientists both are facing a new problem since last decades that how to enhance the tribological properties of a composite. The necessity of metal substitution in radar, sensors, electronic device, microwave absorbers and rocket application requires stunning tribological properties for better working life cycle. There are numerous answers for taking care of the issue. The most widely recognized is to fill the polymer with different powders. [3-5].

Ferrite nanoparticles are critical class of attractive materials from application perspective and are widely utilized as lasting magnets in market in light of their low value, high attractive execution, non-toxicity, biocompatibility, high resistance to corrosion and environment-friendly qualities [6-9].So due to decent electromagnetic properties Ferrite nanoaprticles is one of the most promising and greatest characterized filler particles, which is united with epoxy, i.e., to record media [10] and in medical applications [11]. Numerous polymorphic forms of Ferrite nanoparticles are used in the field of aerospace, sensor and other electronic fields. Epoxy are extensively used in numerous applications due to their higher adhesion, high strength, and resistance to creep, heat and chemicals [12].

But ferrite nanoparticle epoxy composites have some limitations. Ferrite nanoparticle-strengthened properties of epoxy nano-composites but suffered low tribological problems. One of the main reason of low tribological characteristics of ferrite epoxy composites is poor dispersion of ferrite particles into epoxy resin [13]. Main concern in the fabrication of ferrite/polymer composites is to accomplish the good dispersion and interfacial strength between ferrite nanoparticles and the epoxy resin. Actually during the process when ferrite nanoparticles dispersed in to epoxy resin slurry, powerful magnetic attraction between the ferrite nanoparticles started agglomeration of ferrite nanoparticles. Agglomerates of hydrophilic ferrite nanoparticles could not be dispersed homogeneously in organic epoxy normally assisted to establishment of imperfections belongs to micro-structural in the ferrite nanoparticles epoxy composites. This types of imperfection affect the composite's tribological properties [14-16].

In current study point of view, Ferrite particles were surface treated by APS coupling agent. Modified ferrite nanoparticle epoxy were fabricated through dispersion of modified ferrite nanoparticles and normal ferrite nanoparticles into epoxy matrix. For examination the surface treatment effect on ferrite nanoparticles with APS by tribological properties of composite; microstructural characterization was analyzed using XRD and tribological characterizations by Pin on Disc wear test.

2. Experimental

2.1. Materials

The commercially available epoxy (CY230) which is based on DGEBA and hardener (HY 951) were procured from M/S Huntsman India Limited. APS were purchased by Himedia Laboratory Pvt. Ltd. India. Ferrite nanoparticles (FNs) (M.W.159.69, 95%) were purchased from Research-Lab Industries, Mumbai, India. All chemicals and solvents used without further purifications. Deionized water (conductivity10^{-8} S/cm) was used in the preparation.

2.2. Synthesis of modified ferrite nanoparticles epoxies (MFNEs)

The synthesis of MFNEs has been implemented into two phases. Primarily, ferrite nanoparticles (FNs) modified by surface treatment. In this process firstly, 10gm FNs was mixed into 1 liter deionized water in a round bottom glass flask using ultra-sonication for 15 min. The surface modification of ferrite particles were started with the help of 9 ml APS which added drop wise into ferrite deionized slurry. For vigorous stirring a Teflon-coated powerful stirring bar has been used for 12 hour at 45±5 ◦C. The achieved centrifuged modified nanoparticles was cleaned by distilled water- ethanol solution (1:1) four times and then dried at 90±5 ◦C to acquire MFNs and on secondary stage, Epoxy composites was filled with loading different weight percentage of modified ferrite nanoparticle and normal ferrite nanoparticles. Separately, both ferrites particles were firstly mixed properly into resin (CY 230) and then applied ultra-sonication in ice bath for 15 min. The acquired mixture was firstly 1 hour heated at 90 ±10 ◦C in microwave and then cooled down to 45±5 ◦C. After that addition of the curing agent (HY 951), the epoxy- modified ferrite nanoparticle and epoxy - normal ferrite nanoparticles were set aside for curing at normal room temp. for 72 hours in teflon moulds to avoid stacking.

2.3. Characterization

The crystallinity of the MFNEs and FNEs samples were characterized using XRD analysis. The diffraction patterns were recorded by a Rigaku-Geiger-flex diffracto-meter using Cu–Kα radiation at 25◦C. The investigation was performed though"40 mA current, 40 kV voltage and 1.54 A wavelength". The 2θ range was 4–80 degree with a rate of 1degree /min. Tribological tests of MFNEs,

IOP Conf. Series: Materials Science and Engineering **802** (2020) 012007 doi:10.1088/1757-899X/802/1/012007

FNEs and cured epoxy were performed on A pin-on-disk tribometer procured by DUCOM, India. ASTM G 99 standard was used for Specimen's dimension. The tribometer had a hardened alloy steel disc working as a counter body. Each Pin shaped epoxy composite with dimension 20 mm in length and dia10 mm has taken three reading using a pin-on-disc tribometer at two different steady state conditions i.e. 200 RPM and 400 RPM, by keeping other constraints constant.

3. Results and discussion

Figure 1 shows the XRD for MFNs and FNs samples. The found XRD patterns attributed for the "rhom-bohedral phase" for "α-Fe2O3" conferring to ICDD ref. card04-008-7623. The shown XRD patterns exhibit high crystalline nature and pure structure of "α-Fe2O3" specimens [17]. Also, the expansion in heights of the intensity peaks approves enchasing the intensity of crystallization. The average crystalline size of MFNs was found to be 16.48 nm using Scherrer's formula as explained in "Eq. (1)".

$$T = K\lambda/\beta cos\theta \quad (1)$$

Here, "T" (average crystalline size), "K" (0.89), "θ" (Bragg's angle) and "λ" (wavelength) and "β" (FWHM of intense XRD peak at θ).

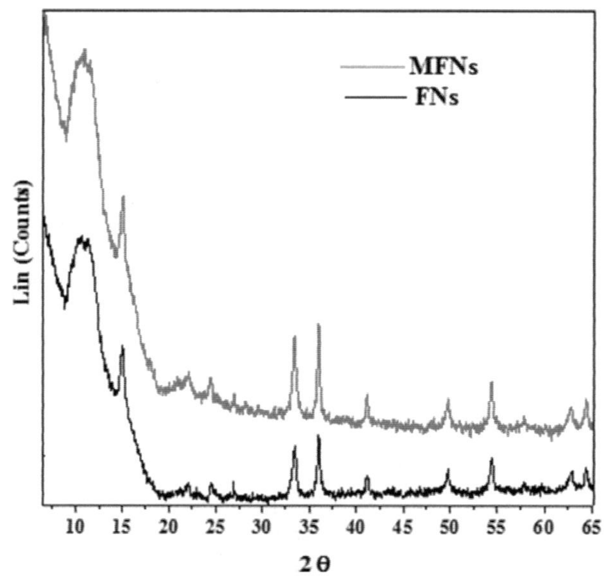

Figure 1. XRD pattern of MFNs and FNs

Figure 2 – Figure 3 shows the variations in different wear rate of cured epoxy, FNEs and MFNEs with different concentration of MFNs (wt%) . Figure 2 represents the variations in wear rate (W_R X10^{-2}, mm^3/m) of cured epoxy and MFNEs with different concentration of MFNs (wt %) at two different rpm 200 and 400 respectively. The variations in decrement in W_R of MFNEs over cured epoxy have been represented in the Figure 2 – Figure 3.The cured epoxy has shown W_R by 3.16 and 3.39 at 200rpm and 400rpm respectively. With increasing concentration of MFNs a notable reduction in W_R has been shown for each of MFNEs. Figure 3 has shown W_R comparison between cured epoxy, ferrite nanoparticle epoxies (FNEs) and modified ferrite nanoparticle epoxies (MFNEs).

At 200 rpm, MFNEs-I shows a decrease in W_R over cured epoxy by 22.15%. Increase in concentration of MFNs to 1wt%, MFNEs-II, has further contributed decrease in W_R of MFNEs over cured epoxy by 34.17% .Similarly at MFNs concentration 1.5wt%, MFNEs-III and MFNs concentration 2wt%, MFNEs-IV shows a decrement in value of W_R by 41.4 and 51.26 respectively over cured epoxy. At 400 rpm, MFNEs-I shows a decrease in W_R over cured epoxy by 15.92%.

Increase in concentration of MFNs to 1wt%, has further contributed decrease in W_R of MFNEs over cured epoxy by 25.07%. Similarly at MFN concentration 1.5wt%, MFNEs-III and MFNs concentration 2wt%, MFNEs-IV shows a decrement in value of W_R by 28.31 and 33.03 respectively over cured epoxy (Figure 2).

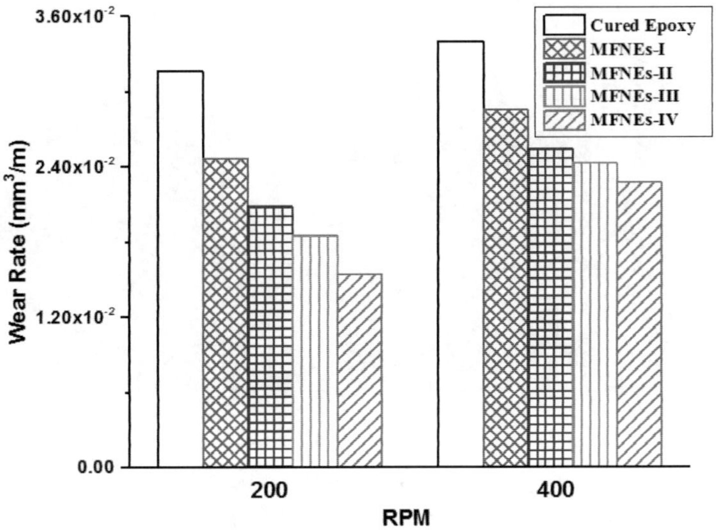

Figure 2. Wear Rate of different MFNEs with varying speed at 200rpm and 400rpm

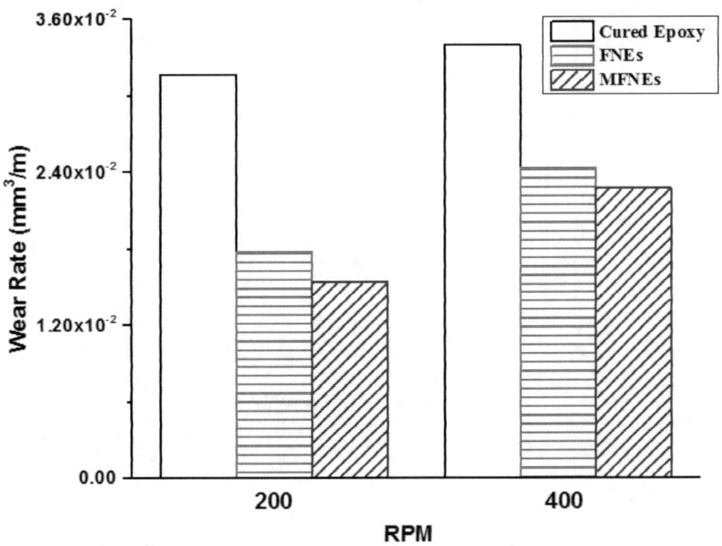

Figure 3. Variation in Wear Rate of Cured Epoxy, FNEs and MFNEs at 200 rpm and 400rpm

In Figure 3 has shown a comparison of W_R which shows that due to good dispersion and enhance adhesion between MFNs and epoxy, lower W_R of MFNEs comparison to FNEs and cured epoxy. The percentage decrease in the W_R of FNEs and MFNEs by 43.98 and 51.26 respectively as compared to cured epoxy at 200rpm and similarly 28.32 and 33.03 respectively as compared to cured epoxy at 400 rpm.

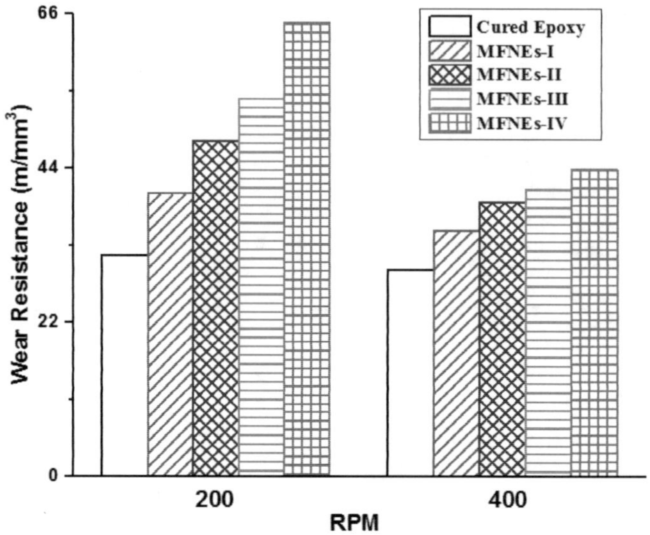

Figure 4. Wear Resistance of different MFNEs with varying speed at 200rpm and 400rpm

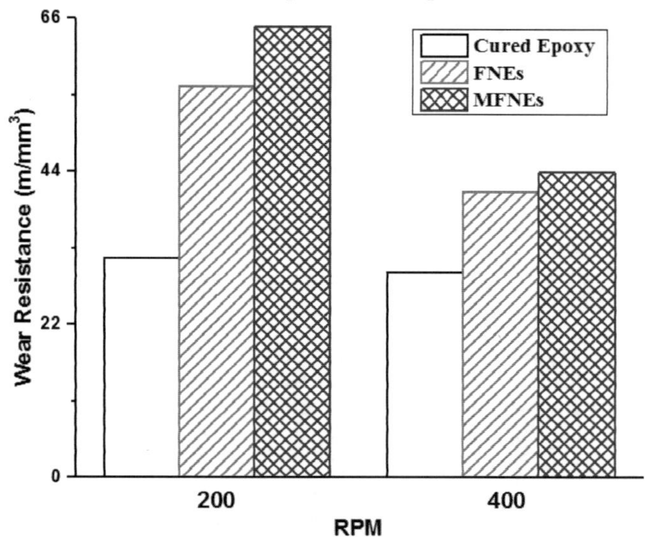

Figure 5. Variation in Wear Resistance of Cured Epoxy, FNEs and MFNEs at 200 rpm and 400rpm

Figure 4 represents the variations in wear resistance (W_{RE} m/mm^3) of MFNEs with different concentration of MFNs (wt %) at two different rpm 200 and 400 respectively. The variations in increment in W_{RE} of MFNEs over cured epoxy have been represented in the Figure 4 – Figure 5.The cured epoxy has shown W_{RE} by 31.6 and 29.45 at 200 rpm and 400 rpm respectively. With increasing concentration of MFNs a notable increase in WRE has been shown for each of MFNEs. Figure 5 has shown W_{RE} comparison between cured epoxy, ferrite nanoparticle epoxies (FNEs) and modified ferrite nanoparticle epoxies (MFNEs).

At 200 rpm, MFNEs-I shows an increase in W_{RE} over cured epoxy by 28.16 %. Increase in concentration of MFNs to 1wt%, MFNEs-II, has further contributed increase in W_{RE} of MFNEs over cured epoxy by 51.89% .Similarly at MFNs concentration 1.5wt%, MFNEs-III and MFNs concentration 2wt%, MFNEs-IV shows an increment in percentage value of W_{RE} by 70.88% and 105.06% respectively over cured epoxy. At 400 rpm, MFNEs-I shows an increase in W_{RE} over cured epoxy by 18.91%. Increase in concentration of MFNs to 1wt%, has further contributed increase in W_{RE} of MFNEs over cured epoxy by 33.34%. Similarly at MFNs concentration 1.5wt%, MFNEs-III and MFNs concentration 2wt%, MFNEs-IV shows an increment in percentage value of W_{RE} by 39.69% and 49.16% respectively over cured epoxy (Figure 4).

In Figure 5 has shown a comparison of W_{RE} which shows that due to good dispersion and enhance adhesion between MFNs and epoxy, higher W_{RE} of MFNEs comparison to FNEs and cured epoxy. The percentage increase in the W_{RE} of FNEs and MFNEs by 78.29 and 105.06 respectively as compared to cured epoxy at 200 rpm and similarly 39.69 and 49.16 respectively as compared to cured epoxy at 400 rpm.

4. Conclusions

The surface treatment of FNs was efficiently accomplished by using APS as coupling agent to achieved MFNs. XRD pattern confirmed that the achieved MFNs has high crystalline structure and their size reduced to 16.48nm. Improved dispersion of MFNs into epoxy resin and enhanced adhesiveness between filler and epoxy resin was gained due to APS coupling. Due to APS surface treatment agglomeration of ferrite particle has been reduced and ferrite particles homogeneously dispersed in epoxy resin which has been shown better tribological behaviour of ferrite nanoparticles epoxy composites compare to normal ferrite nanoparticles epoxy composite. Results has been Shown that percentage decrease in the wear rate of FNEs and MFNEs by 43.98 and 51.26 respectively as compared to cured epoxy at 200rpm and similarly 28.32 and 33.03 respectively as compared to cured epoxy at 400 rpm whereas percentage increase in the wear resistance of FNEs and MFNEs by 78.29 and 105.06 respectively as compared to cured epoxy at 200 rpm and similarly 39.69 and 49.16 respectively as compared to cured epoxy at 400 rpm.

References

[1] Stabik J, Chrobak A, Haneczok G and Dybowska A 2011 Magnetic properties of polymer matrix composites filled with ferrite powders *Archives of Material Science and Engineering* **48(2)** 97-102.

[2] Hsiang H, Chen W S, Chang Y L, Hsu F C and Yen F S 2011 Ferrite load effects on the mechanical and electromagnetic properties of nizn ferrite powders-epoxy resin coatings *American Journal of Materials Science* **1(1)** 40-44.

[3] Bîrsan I G , Bria A C V. and Ungureanu V 2009 Tribological and Electrical Properties of Filled Epoxy Reinforced Composites *Tribology in industry* **31(1&2)** 33-36.

[4] Zhang X, Yan X, Guo J, Liu Z, Jiang D and He Q 2015 Polypyrrole doped epoxy resin nanocomposites withenhanced mechanical properties and reduced flammability *J Mater Chem C* **3(1)** 162–76.

[5] Soghieh F J, Maleki B and Behniafar H 2019 Effect of dendrimer-functionalized magnetic iron oxide nanoparticles on improving thermal and mechanical properties of DGEBA/IPD epoxy networks *High Performance Polymers* **31(1)** 1–8.

[6] Ling G, He J 2004 The influence of nano-Al_2O_3 additive on the adhesion between enamel and steel substrate *Material Science Engineering A Structure* **379** 432–436.

[7] Giannelis E 1996 Polymer layered silicate nanocomposites *Advance Materials* **8** 29–35.

[8] Guo J, Zhang X, Guo H, Wang Y, Yan X, Ding D, Long J, Tadakamalla S, Wang Q, Mojammel A K, Jingjing L, Zhang X, Brandon L W, Sun L, David P Y and Suying W Z G 2014 Reinforced magnetic epoxy nanocomposites with conductive polypyrrole nanocoating on nanomagnetite as a coupling agent *RSC Adv.* **4** 36560–72.

[9] Sun T, Fan H, Liu X Z and Zhanjunwu 2015 Fe2O3-epoxy composite with enhanced mechanical properties *Materials and Design* **87** 10–16.

[10] Fulco A P, Melo J D, Paskocimas C A, Medeiros S N , Machado F L and Rodrigues A R 2015 Magnetic properties of polymer matrix composites with embedded ferrite particles *NDT and E International* **77** 42-48.

[11] Hora D, Lednicky F E, Petrovsky E and Ka A K 2004 Magnetic characteristics of ferrimagnetic microspheres prepared by dispersion polymerization, *Macromol. Mater. Eng.* **289** 341–348.

[12] Naguib H M, Ahmed M A , Shanab Z L A 2017 Silane coupling agent for enhanced epoxy-iron oxide nano composite *Journal of Materials research and Technology* **7(1)** 21-28.

[13] Guo J, Ye X, Liu W, Wu Q, Shen H and Shu K 2009 Preparation and characterization of poly (acrylonitrile-co-acrylic acid) nanofibrous composites with Fe3O4 magnetic nanoparticles *Materials Letters* **63** 1326–28.

[14] Zaidi M G H , Bhullar N, Sharma D, Agarwal V, Alam S, Rai A K and Pant R P 2007 Synthesis of polyvinyl pyridine ferrite nanocomposites in supercritical carbon dioxide *Journal of Nanostructured Polymers and Nanocomposites* 103-109.

[15] Kalendova A, Vesely D, Kohl M and Stejskal J 2014 Effect of surface treatment of pigment particles with polypyrrole and polyaniline phosphate on their corrosion inhibiting properties in organic coatings *Progr. Org. Coat* **77(9)** 1465–83.

[16] Kim H J, Jung D H , Jung I H , Cifuentes J I , Rhee K Y and Hui D 2012 Enhancement of mechanical properties of aluminium/epoxy composites with silane functionalization of aluminium powder *Composites B:Eng.* **43** 1743– 48.

[17] Zheng H, Cheng Y, Wang S, Bao F, Zhou H and Wei F 2006 QuasicubicFe2O3 nanoparticles with excellent catalytic performance *J. Phys. Chem. B* **110** 3093–7.

[18] Zaidi M G H, Sah P L, Alam S and Rai A K 2009 Synthesis of epoxy ferrite nanocomposites in supercritical carbon dioxide *Journal of Experimental Nanoscience* **4** 55–66.

[19] Chen W S, Chang Y L , Hsiang H I, Hsu F I and Yen F U 2011 Effects of titanate coupling agent on the dielectric properties of NiZn ferrite powders–epoxy resin coatings *Ceramics International* **37** 2347–52.

IOP Conf. Series: Materials Science and Engineering **802** (2020) 012008 doi:10.1088/1757-899X/802/1/012008

Comparison of a Classical Cyclone Separator and Protruding Surface Cyclone Separator using CFD Software

Vinay Sati[1a*], Shivasheesh Kaushik[1a*], Dr. Rahul Kshetri[2a], Dr. Kuldeep Panwar[2b], Rahul Pandey[2c]

[1a*]Assistant Professor, Department of Mechanical Engineering, Amrapali Group of Institute, Haldwani, Uttarakhand, India.

[1a]Assistant Professor, Department of Mechanical Engineering, Shivalik College of Engineering, Dehradun, Uttarakhand, India.

[2c] Scholar, B.Tech, Department of Mechanical Engineering, Amrapali Group of Institute, Haldwani, Uttarakhand, India.

[2a] Department of Production and Industrial Engineering, G.B. Pant University of Agriculture & Technology, Pantnagar, Uttarakhand, India.

[2b]Associate Professor, Department of Mechanical & Automation Engineering, Delhi Technical Campus, Greater Noida, India.

*Corresponding author email: vinaysati379@gmail.com

Abstract. Industrial smoke contains various hazardous particles, ash namely to be one of the various substances. The cyclone separator is a specially designed device used to separate the ash particles from the stream of air. The stream of dirt and air mixture flows in two vortexes, the inner vortex and the outer vortex, this device in turns acts as a scrubber and the ash particles get collected down in the ash collector whereas the fresh air stream is released by the device. To increase the efficiency, i.e. to increase the ash collecting capacity of the device, we have tried to compare a classical cyclone separator with the one we created which has protruding surfaces internally and the comparative analysis is done using the CFD software ANSYS R16.0.

1. Introduction

1.1. CFBC Technology:

Cyclones are generally used with CFBC (Circulating Fluidized Bed Combustion) type of boilers, the coal is burnt in a bed comprising of hot particles and maintaining the flow of air through the bed. At sufficiently high air speeds, the bed will act as a fluid resulting in quick combination of the particles. This fluidised action will promote the complete combustion of coal at a relatively low temperature and will provide a channel to transfer heat generated efficiently from the bed to the steam tubes.

The boiler in which coal is burnt in an environment of high concentration of bed material (mineral matter) is derived from combustion of coal retained by using cyclone. This bed material is fluidized by primary air which is a part of combustion air.

1.1. Cyclone Separator:

Relatively coarse particle of limestone and unburnt char are captured in the cyclone and are recycled back near the base of the furnace. Finer solid residues (ash and spent sorbents) are generated during the process of combustion and desulphurization which leave the furnace, escaping through the cyclones, but they are collected by a bag-house or electrostatic precipitator which is further located downstream. A cyclone is a simple device which provides a sufficiently high degree of separation for a minimum pressure drop. Most of the solids (approx. 98%) are successfully removed by a cyclone separator when operating at designed flow rates. Cyclone Separators offer removal of particulate matter from air streams at very economical cost of operation and maintenance. A cyclone separator consists of mainly two parts namely, an upper cylindrical part referred to as the barrel and a lower conical part referred to as cone, also known as hopper. The stream of air enters tangentially at the top of the barrel from the inlet and travels downward by the action of gravitational force into the cone forming an outer vortex. Increasing air velocity in the outer vortex causes centrifugal force on the particles, separating them from the air stream. When the air reaches the bottom ofthe cone, an inner vortex is created in the reverse direction and exits from the top as clean air (free of charge), while the particulates are collected at the bottom of the cyclone. The separation efficiency of the separators depends on the size of the ash particles being separated as shown in figure number 2.

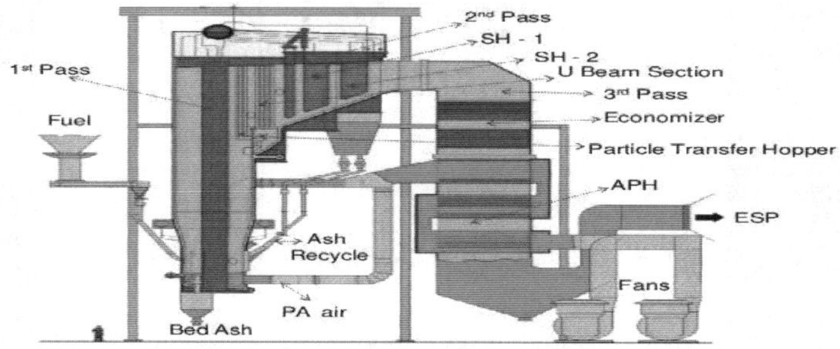

Figure Number 1. A CFBC type boiler arrangement.

2. Literature Review

CFD acronym for Computational Fluid Dynamics has evolved over the past few years which enables designers to design customised piping arrangement for a required system as per the requirements organisation specifications. Although, CFD offers results with great precision and accuracy, but in general practice the numerical values differ by a certain factor as of various external factors are present in reality. The difference in values is due to CFD being done in an ideal case solution. Most CFD solve the Navier-Stokes equation either in the Lagrangian or Eulerian approach. Apart from these methods, some solve the Boltzmann equation instead of Navier-Stokes equation. CFD can broadly be classified under two categories; namely conventional and accelerated methods.

Accelerated methods are further classified as advanced numerical methods and hardware methods. The advanced numerical methods hierarchy which is explained as hardware methods is beyond the scope of this paper. Advanced numerical methods are further classified into three categories as: -

- Mesh based methods
- Mesh free methods
- Hybrid methods

Out of these three methods, the most commonly employed is the mesh-based methods in which there are certain meshed structure created on the geometry of the body. The mesh designs can vary from very fine complex meshes to simple larger meshes; as applicable by the need. Some of the common meshed based methods are listed as: -

- Reduced Order Modelling (ROM)
- Marker and Cell (MAC)

In the mesh free methods, direct surface nomenclature is done by determining the direction of flow and the corresponding parametric values are applied. This is generally used as a macroscopic approach as surface analysis is done in this case. But, in Mesh-based model the microscopic analysis is done. Mesh offers variation of parameters at every instant of the mesh, which depends on how much fine meshing has been done. For in-depth analysis of flow, Mesh-based models are preferred as compared to Mesh free models. The advantage of Mesh free model is that it can provide quick results; which translates into easy adaptation in industries wherein continuous production of goods is taking place. The mesh free methods generally employed are as follows: -

- Smoothed Particle Hydrodynamics (SPH)

- Fast Multipole Method (FMM)
- Method of Fundamental Solutions (MFS)
- Finite Pointset Method (FPM)

Moving Particle Semi-Implicit Method (MPS)Our research has been carried out by the conventional methods as they offer a great degree of accuracy, although being slow as compared to accelerated methods. The most popular conventional methods used in practice are: -

- Finite Element Method (FEM)
- Finite Volume Method (FVM)
- Finite Difference Method (FDM)

These methods solve the Navier-Stokes equation, which are the governing equations of the CFD Software. These equations describe the conservation of mass, momentum and energy. The method solves equation of complexity order n^3,where n is the degree of freedom. To solve the program in stages, concepts of dynamic programming can be used.

3. Research Methodology

The research was carried out from scratch and performed in multiple stages. The primary stage of the research involved the designing of the cyclone models, which was done with the use of Solidworks 2016 software.

3.1. First Stage:

We started with the designs that needed to be checked for their integrity. This was again done in the same software by applying the geometric region tests. Since all parts of the models were perfect regions as obtained from the tests performed on them; the first step in the research process was completed as the geometric bodies had perfect regions indicating that there shall be no losses due to leakage anywhere when we perform the fluid flow simulation.

3.2. Second Stage:

In the second stage of research, we imported the design files into the CFD software ANSYS R16.0. Then, we assigned the names to the sections of the pipe geometry as inlet, body and outlet. This named selection was done so that analysis at every region of the pipe as a whole body can be performed. This was followed by fine meshing of each section was done in order to provide accurate graphs while simulation is done. Lastly in this stage, we applied the flow direction to the body as inlet>body>outlet respectively.

3.3. Third Stage:

The third stage of the research was to select the applicable equations for the meshed model generated. As mentioned, we applied the Cyclone equation and selected the K-Epsilon model to obtain the results for our geometry. These equations have been explained further in the mathematical modelling section.

3.4. Fourth Stage:

In the fourth stage, which being the most important stage of the research was to do the simulation of flow. Now using FLUENT we applied the fluid flow through the geometry designed one by one, for each of the case we obtained the velocity and the pressure contours which are illustrated as the variation of the velocity and the pressure component at each point. Also, the graphs obtained indicated the variation of velocity and the pressure within the pipe.

Multiple number of simulations were done to check the integrity of the results obtained. We, then concluded that the results obtained from the simulation was genuine and hence our research work came to its conclusion. The results obtained are of for ideal conditions and have not been practically verified. There is a possibility of a variation in the values mentioned in the paper by a certain factor.

4. Research Gap

These results were obtained by CFD software are under ideal conditions with standard wall functions. Henceforth, during practical validity we may employ material depending on various factors, some of them being GCI, PVC, SS of varying grades with each possessing different chemical properties. These may affect the values of fluid parameters when tested under these materials. Also, various losses like losses due to friction will be taken into account when practically performed. Various environmental conditions prevailing may also affect the calculations in general practice as this was an ideal case. Due to lack of equipment andcertain research tools the practical validity of the models discussed could not be tested and hence comparison between the practical and ideal approach could not be done.

5. Assumptions made

- Gas density is negligible as compared to the density of particles
- Gravitational forces are negligible as compared to the centrifugal forces
- Tangential velocity (V_t) does not vary in the axial direction, it is neglected due to short distance being covered; thus, relative velocity is purely radial
- Particles are spherical of low size and with small enough relative velocity to apply stokes law and not disturb the natural flow around it.
- No interaction between the particles.

6. Mathematical Modelling

There are two models used for the analysis- one is the simple cyclone (SC-1) and the other one is protruding cyclone (PC-2) as shown in figure number 2 and 3

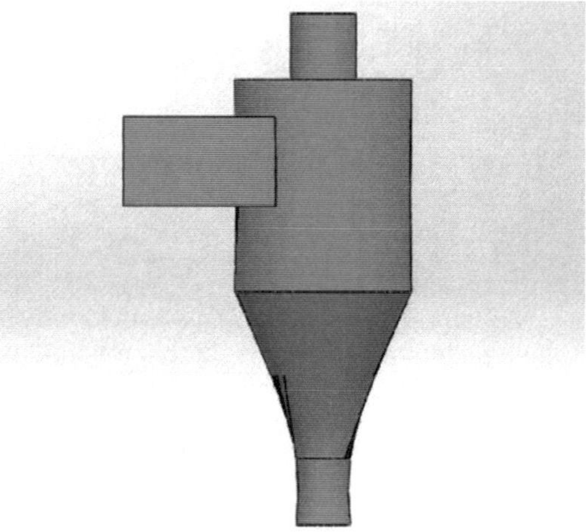

Figure Number 2. The SC-1 solid geometry.

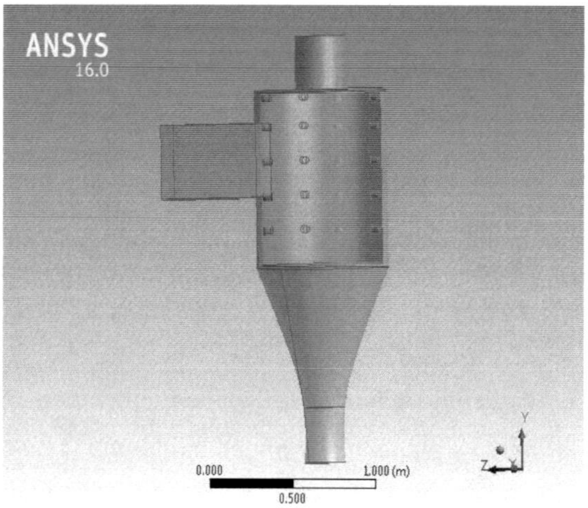

Figure Number 3. The PC-1 solid geometry.

The details about SC-1 are as follows: -
Mass properties of simple cyclone (Part Configuration - Default)
Output coordinate System: -- default --
Density = 7800.00 kilograms per cubic meter
Mass = 5617.28 kilograms
Volume = 0.72 cubic meters
Surface area = 5.41 square meters
Centre of mass: (meters)
X = 0.02
Y = 1.37
Z = 0.05
Principal axes of inertia and principal moments of inertia: (kilograms * square meters)
Taken at the centre of mass.
Ix = (0.03, 0.99, 0.10) Px = 529.46
Iy = (0.31, -0.10, 0.95) Py = 1277.03

IOP Conf. Series: Materials Science and Engineering **802** (2020) 012008 doi:10.1088/1757-899X/802/1/012008

Iz = (0.95, -0.00, -0.31) Pz = 1448.99
Moments of inertia: (kilograms * square meters)
Taken at the center of mass and aligned with the output coordinate system.
Lxx = 1431.88Lxy = 25.12 Lxz = 52.76Lyx = 25.12 Lyy = 537.35 Lyz = 72.19
Lzx = 52.76 Lzy = 72.19 Lzz = 1286.24
Moments of inertia: (kilograms * square meters)
Taken at the output coordinate system.
Ixx = 11984.49 Ixy = 155.51 Ixz = 57.39
Iyx = 155.51 Iyy = 552.28 Iyz = 446.74
Izx = 57.39 Izy = 446.74 Izz = 11827.16

The only difference between the two models was of the studs, as the SC-1 has no studs, i.e.it has no protruding surfaces at the inner wall of the separator whereas the PC-2 has studs on its surface.
The dimension of the studs is: -
Diameter= 40 mm
length= 50 mm

7. Data Reduction

- $F_d = -6\pi r_p \mu V_r.$
- $F_c = m\frac{V_t^2}{r}$
- $F_b = -V_p \rho_f \frac{V_t^2}{r}$
- $m\frac{dV_r}{dt} = F_d + F_c + F_b$

For the sake of simplification of these above equations, we consider that the particle under observation has reached "terminal velocity", i.e., there is no component of acceleration. This kind of situation occurs in the case when the radial velocity has generated enough drag force to counter the acting centrifugal and buoyancy forces.

$$F_d + F_c + F_b = 0$$

$$-6\pi r_p \mu V_r + \frac{4}{3}\pi r_p^3 \frac{V_t^2}{r}\rho_p - \frac{4}{3}\pi r_p^3 \frac{V_t^2}{r}\rho_f = 0$$

In non-equilibrium conditions when the component of radial acceleration is not zero, then the general equation must be considered. Rearranging the terms, we get-

$$\frac{dV_r}{dt} + \frac{9}{2}\frac{\mu}{\rho_p r_p^2}V_r - \left(1 - \frac{\rho_f}{\rho_p}\right)\frac{V_t^2}{r} = 0$$

Experimentally it has been found that the velocity component of rotational flow is directly proportional to the square of the radius(r^2), therefore

$$V_t \propto r^2.$$

8. Input Parameters

Analysis of the flow conditions was done using ANSYS Fluent 16.0 CFD Software. These equations are solved by converting complex partial equation into simple algebraic equation. All the geometries SC-1, PC-2 are 3-D rigid solids comprising of 1 region 3 named selections as inlet, outlet, and path. These 3-D geometries were used for solving momentum and energy equations. The initial phase velocity of the flow was defined at the inlet section of the pipe upstream. Acceleration due to gravity has been taken as 9.81 m^2/sec. The standard wall functions were applied to the κ-ε turbulence models for solving the problems. The initial velocity given to all the geometries were 3 m/sec.

9. Results and Observations

The charts of the pressure and velocity indicating variation in their respective values were obtained as shown in figure number 4, 5, 6 and 7 also the velocity and pressure contours obtained are as shown below: -

IOP Conf. Series: Materials Science and Engineering **802** (2020) 012008 doi:10.1088/1757-899X/802/1/012008

9.1. Variation of Pressure along Cross Section:

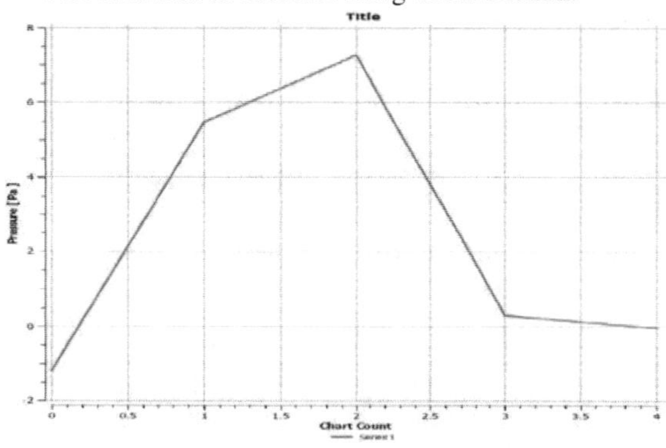

Figure Number 4. Pressure observed in SC-1

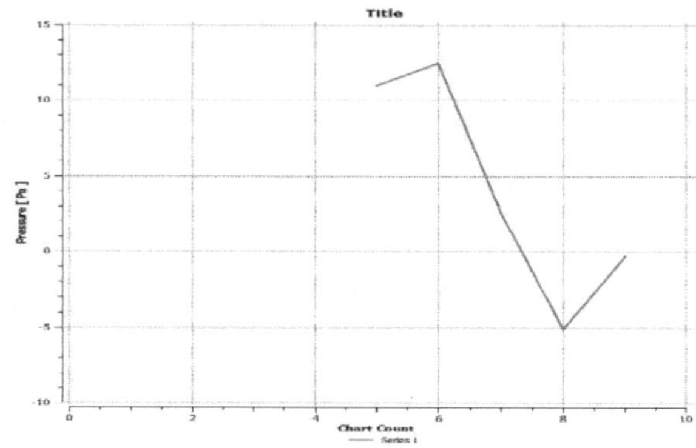

Figure Number 5. Pressure observed in PC-2

9.2. Variation of Velocity along Cross Section:

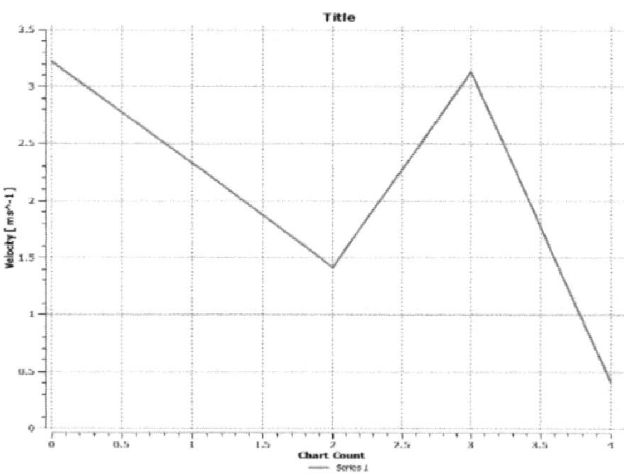

Figure Number 6. Velocity observed in SC-1

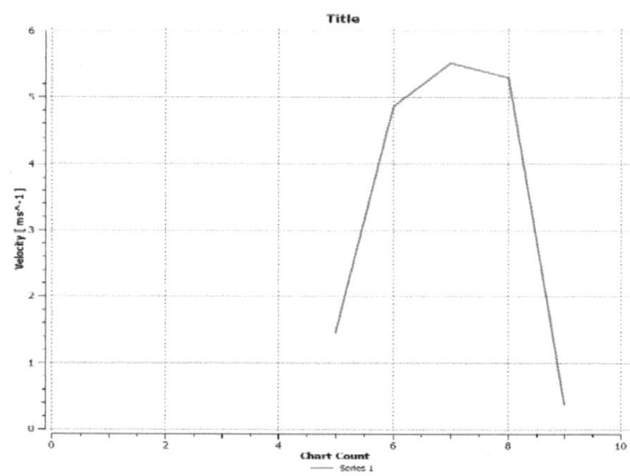

Figure Number 7. Velocity observed in PC-2

9.3. Vortex Core Region of Cyclone Separator:

Figure Number 8. Vortex Core observed in SC-1

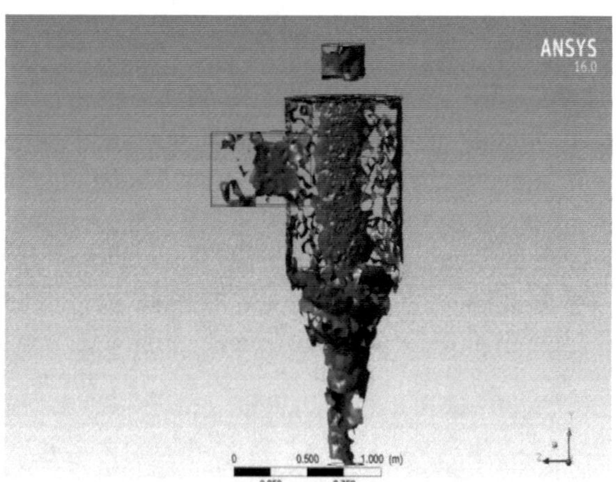

Figure Number 9. Vortex Core observed in PC-2

IOP Conf. Series: Materials Science and Engineering **802** (2020) 012008 doi:10.1088/1757-899X/802/1/012008

9.4. Velocity Contour of Cyclone Separator:

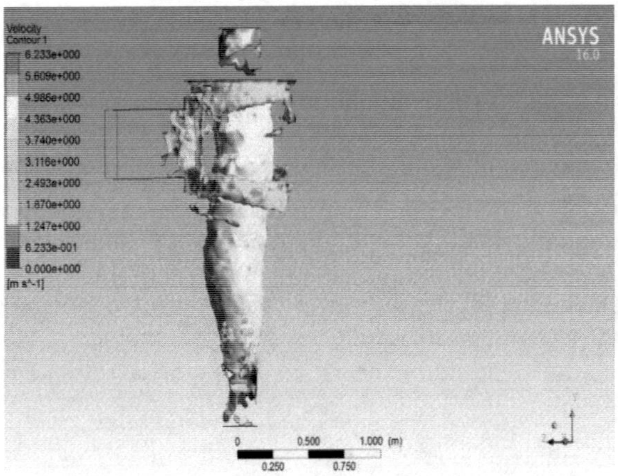

Figure Number 10. Velocity Contour of SC-1

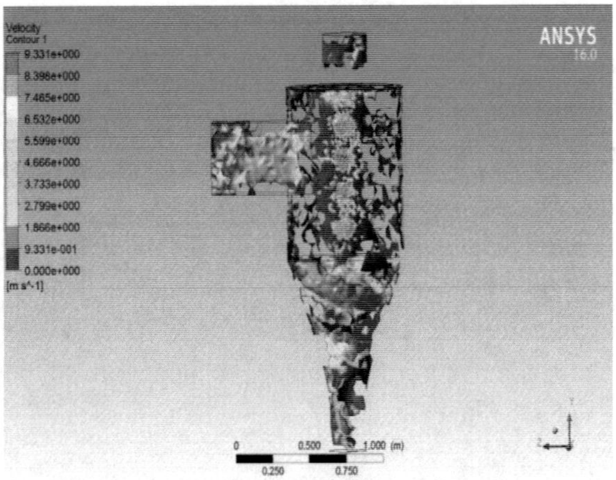

Figure Number 11. Velocity Contour of PC-2

9.5. Pressure Contour of Cyclone Separator:

Figure Number 12. Pressure Contour of SC-1

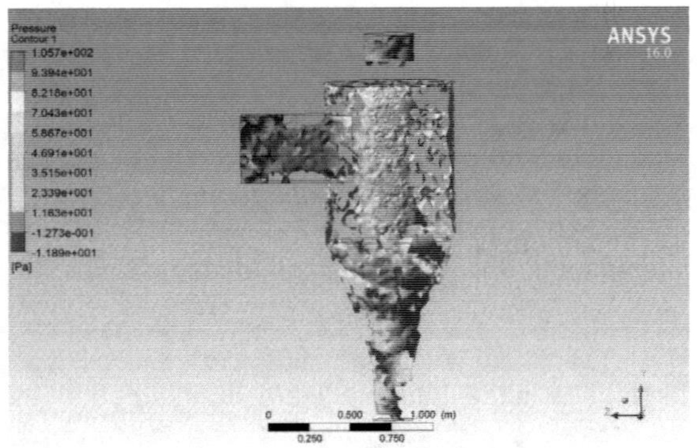

Figure Number 13. Pressure Contour of PC-2

9.6. Velocity Vector of Cyclone Separator:

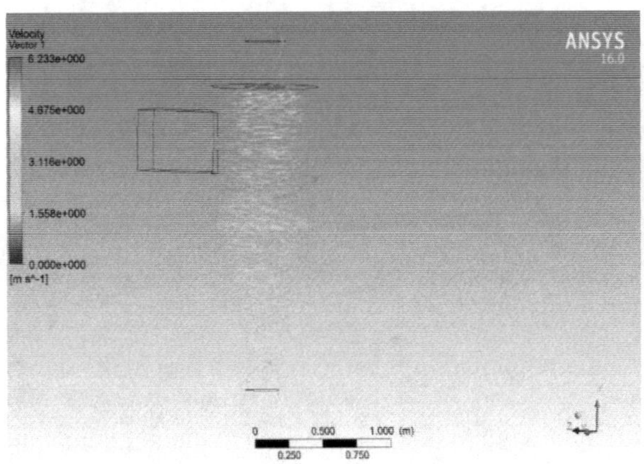

Figure Number 14. Velocity Vector of SC-1

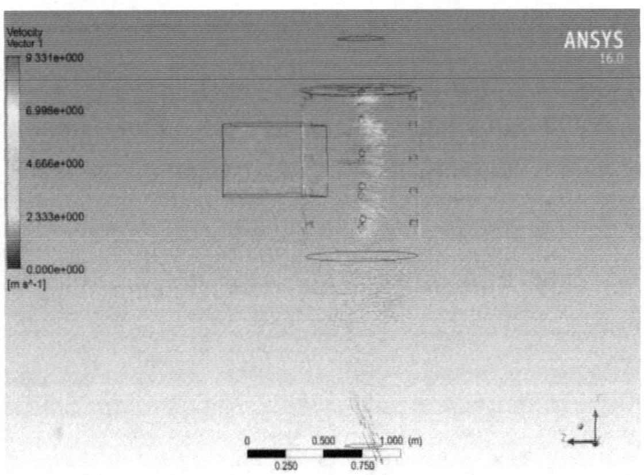

Figure Number 15. Velocity Vector of PC-2

The various pressure zones are indicated in the protruding cyclone clearly depicts that the fresh air free of charge flows out with a pressure of 5.34 Pa.The 3.897E+001 Pa zone indicates the mass flow rate at the inlet of the cyclone; it is evident from the contour that the pressure at the hopper is 3.010 Pa and same is the pressure of the fresh air free of charge coming out of the cyclone as shown in figure number 12 and 13.

The velocity remains in the zone of 3.11 m/s of the fresh charge free air flowing through the device as shown in figure number 14 and 15.

1. Conclusion

Based on the above plots and graphs obtained we can infer that the protruding surface cyclone separator offers better ash collection as compared to the simple cyclone separator; which in turn increases its efficiency with minimal design alteration.This in turn confers that in industrial applications where we deal with CFBC type of boilers we may use protruding surface type cyclone separator to enhance the ash collection and make it more economical for the operations and maintenance as compared to the classical cyclone separator.

Reference

[1] Parnell, C. B. Jr. 1996. Cyclone design for air pollution abatement associated with agricultural operations. In Proc. Beltwide Cotton Production Conferences.

[2] K.S. Lim, H.S. Kim, K.W. Lee, 2004,Characteristics of the collection efficiency for a cyclone with different vortex finder shapes, J. Aerosol Sci. 35, 743–754.

[3] W.E. Conard, H.G. Conard, T.Szylowlee, Insert for a cyclone separator, US, 0008072A1, 2002.

[4] J. Casal and J.M. Martinez-Benet. A better way to calculate cyclone Pressure Drop. Chemical Engineering, 90(2):99100,1983

[5] B.Wang, D.L. Xu, K.W. Chu, and A.B. Yu. 2006, Numerical study of gas-solid flow in separator.Applied Mathematical Modeling, 30:1326-1342.

[6] D. Benoni, C.L. Briens, T. Baron, E. Duchesne and T.M. Knowlton, 1994, "A procedure to determine particle agglomeration in a fluidized bed and its effect on entrainment", Powder Technology, 78, 33-42.

[7] Barth W. 1956, Design and layout of the cyclone separator on the basis of new investigations. Brennstoff-Warme-Kraft 8: 1-9

[8] P. Sivakumar, 2015, Development of an optimum CFBC Cyclone separator with REPDS for low pressure drop and denudation rates using CFD, International Journal of Applied Engineering Research ISSN 0973-4562 Volume 10, Number 13 (2015).

IOP Conf. Series: Materials Science and Engineering **802** (2020) 012009 doi:10.1088/1757-899X/802/1/012009

Comparative Study for Thermal and Fluid Flow Peculiarities in Cascading Spiral Inner Tube Heat Exchanger with or without Diverse Inserts over Spiral Tube

Shivasheesh Kaushik[1a*], Dr. Satyendra Singh[1a], Nikhil Kanojia[1b], Kuldeep Rawat[2a], Dr. Kuldeep Panwar[2b]

[1a*, 1b]Assistant Professor, Department of Mechanical Engineering, Shivalik College of Engineering, Dehradun, Uttarakhand, India.
[2a]Department of Mechanical Engineering, School of Engineering & Technology, Shri Venkateshwara University, U.P. India.
[1a]Professor, Department of Mechanical Engineering, B.T.K.I.T. Dwarahat, Uttarakhand, India.
[2b]Associate Professor, Department of Mechanical & Automation Engineering, Delhi Technical Campus, Greater Noida, India.
* Corresponding author email: shivasheeshkecua@gmail.com

Abstract. Present work deals with experimental analysis in which various observation takes place related to performance characteristics like HTR, effectiveness and fluid flow peculiarities like Nu, Re, f, NTU, TPF etc in a cascaded spiral concentric tube type heat exchanger consisting of a GI shell and a cascaded spiral copper inner tube having conical shape inserts, hemispherical insert and plane profile unit for parallel flow. Five spiral cascades with three turns in each are used for the inner side copper tube unit. Water enters at 304°K from the innermost turn along the cascaded spiral tube, and leaves at the outermost turn. The hot fluid inlet for heat exchanger is from bottom of one end of the shell at 333°K, flowing radically across the cascades, and then leaving the exchanger from the top of another end. This paper based on comparative experimental study of the behavior of such system for parallel flow with different shape of inserts, and then compared with the simulation results for same. In this analysis comparative approach is considered for evaluating optimum fluid flow and thermal properties on the basis of geometrical parameters. The results predicted that out of three profiles of insert, plain, conical and half cut elliptical profile, which located over inner copper tube of exchanger, half cut elliptical profile insert shows best results on fluid flow and thermal peculiarities, for mass flow rate range 0.012 to 0.049 Kg/s and Reynolds number range 4236 to 18540, heat transfer rate varies from 1128.06 to 2374.77, Nusselt number varies from 75.15 to 144.05, Friction factor varies from 0.00376 to 0.00305 and effectiveness varies from 0.77 to 0.40 and Number of transfer unit varies from 2.08 to 0.68 for half cut elliptical profile.

1. Introduction

As per previous literature it has been observed that phenomenon of heat transfer plays an crucial role in day to day appliances such as light weight equipments like laptops for effective cooling purpose, heavy components used in industrial sectors like heat exchangers for waste heat recovery, solar based equipments like solar air heater, solar water heater etc regarding effective utilization of non conventional energy resource like sun, that why research e have keen interest on such type of investigation in which effect of various types of insert, micro channel, baffles etc[11,12,13,14,15,18], along with diverse fluid [16,17] on thermal and fluid flow peculiarities involved and responsible for thermal performance enhancement in solar based heat transfer or exchangers type devices. Also, present ongoing researches and demands, it is necessary to find the behavior of differently augmented heat exchangers and heat transfer surface due to several engineering implementation and insinuation dealt with it. More efficacious and veracious heat exchanger type devices can be made through augmentation and economic reconsideration regarding heat transfer. A decrement in heat exchanger

IOP Conf. Series: Materials Science and Engineering **802** (2020) 012009 doi:10.1088/1757-899X/802/1/012009

size has been seen, with its increased performance [1], poor performance of tube type heat exchangers have been noticed while handling high viscosity fluids in laminar flow due to very little heat transfer because of inefficient thermally boundary layer. Control of heat transfer is governed by the extent of flourished boundary layer and thermal conductivity associated with tube material. The effect of curved tube within a CTHE has been analyzed by many researchers in near future. Garimella et al. [2] investigated forced convection heat transfer and determined average heat transfer coefficients of transition flows and laminar flow for coiled type annular ducts. Prabhanjan et al. [3] studied HCHE and compared its result with straight tube specifically for heat transfer rate. Ho et al. [4–6] derived the correlations of HTC for tube-side and air-side. Bhola et al. [7-8]conducted simulation for finding the effect of rectangular insert in parallel flow and counter flow arrangements of CTHE. Analyzing above literatures shows that either spirals, helical or inserts have been used to maximize the heat transfer of a CTHE, which enables us to use a cascaded system for our research.

Nomenclature

Q_d	discharge	v_h	velocity of hot fluid
Q_c	heat transfer rate of hot water	v_c	velocity of cold fluid
m_c	heat transfer rate of cold water	ρ_h	density of hot fluid
c_{pc}	mass flow rate of hot water	ρ_h	density of cold fluid
k_h	mass flow rate of hot water	CTHE	concentric tube heat exchanger
Q_d	specific heat of hot water	HCHE	helical coil heat exchanger
Q_c	specific heat of cold water	HTC	heat transfer coefficient
m_c	heat transfer coefficient	HTR	heat transfer rate
c_{pc}	thermal conductivity of hot water	CSTHE	concentric spiral tube heat exchanger
k_h	thermal conductivity of cold water	CCSTHE	concentric cascaded spiral tube heat exchanger
Q	heat flux		
μ_h	viscosity of hot fluid	CCSICTHE	concentric cascaded spiral inner copper tube heat exchanger
μ_c	viscosity of cold fluid		
		η	thermal performance factor

2. Governing Equations:

The governing equations for investigating the thermal and fluid flow peculiarities in the experimental test set up was carried out by following steps, in first step discharge and velocity is evaluated experimentally for determining the mass flow rate of hot and cold fluids, the governing equation is given as,

$$Q_d = V/t \quad (1)$$
$$Q_d = A * V \quad (2)$$
$$m = \rho * A * V \quad (3)$$

Here, the term m is the mass flow rate of hot and cold fluid which is the major term used to evaluate heat transfer rate of hot fluid and cold fluid which is second step. Q is calculated using the equation given below.

$$Q_h = m_h * c_{ph} (T_{in} - t_{out}) \quad (4)$$
$$Q_c = m_c * c_{pc} (T_{out} - t_{in}) \quad (5)$$

Where, T_{in}, T_{out} and t_{in}, t_{out} is the inlet temperature of shell side heat exchanger, outlet temperature of shell side heat exchanger and inlet temperature of spiral tube side heat exchanger, outlet temperature of spiral tube side heat exchanger.

The third step using LMTD method for evaluating HTC and OHTC which is governed by the following equation:

$Q = h*A*\Delta T$ (6)
$\Delta T = (\Delta T_1 - \Delta T_2)/\ln(\Delta T_1/\Delta T_2 \Delta)$ (7)
$\Delta T_1 = ((T_{out} - t_{in})$ (8)
$\Delta T_2 = (T_{in} - t_{out})$ (9)
$U = 1/(1/h_i + 1/h_o)$ (10)

After determining above values forth step is used to evaluate the dimensionless properties of fluid like Nusselt number, Reynolds number and Friction factor that was governed by following equation.

$Nu = h*D/k = 0.023\, Re^{0.8} Pr^{0.4}$ (11)
The value of Reynolds number is obtained by:
$Re = \rho.V.D/\mu$ (12)
The value of friction factor is obtained by:
$f = \Delta P / (L/D)*(\rho*v^2/2)$ (13)

At last after determining above values for dimensional less parameters fifth step is used to evaluate the effectiveness and number of transfer units through NTU method of CSTHE where, effectiveness is the ratio of actual to maximum possible heat transfer that was governed by following equation.

$\mathcal{E} = Qactual/Qmax$ (14)
$Q = m_c*c_{pc}(t_{out} - t_{in}) = m_h*c_{ph}(T_{in} - T_{out})$ (15)
$Q_{max} = C_{min}*(T_{in} - t_{in})$ (16)
$T_{out} = T_{in} - Q_{max}/C_{max}$ (17)
$t_{out} = t_{in} + Q_{max}/C_{min}$ (18)
$NTU = U*A_s/C_{min}$ (19)
Where, $C_{min} = m*c_p$
$NTU_{parallel} = - [\ln(1-\mathcal{E}(1+c))/(1+c)]$ (20)
$\eta = (Nu/Nu_s)/(f/f_s)$^0.33 (21)

3. Experimental Test Set Up Layout and Instrumentation:

Figure 3.1 shows a simplistic diagram of experimental set up and Figure 3.2 shows the simplistic diagram of plane geometrical profile of spiral tube and spiral tube with different geometrical shapes inserts like triangular insert, hemispherical insert located over spiral tube. The test rig consists of a CTHE with cascaded spiral inner tube and data procuring system. Water is the active fluid. Whole set up is fabricated in such a way that its components can be altered or overhauled easily. In addition to the loop component lively set of apparatus for measuring and controlling the temperature and flow rate of all fluids is installed at all important points in the circuit. In the experimental set up six thermocouples T1, T2, T3, T4, T5 and T6 are installed at inlet and outlet of shell side GI pipe, inlet and out let section of spiral copper tube, hot water tank and cold water tank for evaluating the fluid temperature of hot water and cold water, hot fluid moves from hot water tank to inlet section of shell side GI Pipe to outlet section and cold fluid moves from cold water tank to inlet section of spiral copper tube side to outlet section. Flow rate measurement devices were also installed at inlet and out let section of tube and pipe for determining the pressure drop and mass flow rate. During experimentation readings were noted with the help of flow meters, pressure meters and digital temperature indicator device where all thermocouple are attached. Table number 3.1 shows all the geometrical dimensions of the cascaded spiral tube heat exchanger which were used to fabricate the set up and Table number 3.2 shows all the operating condition used during experimentation.

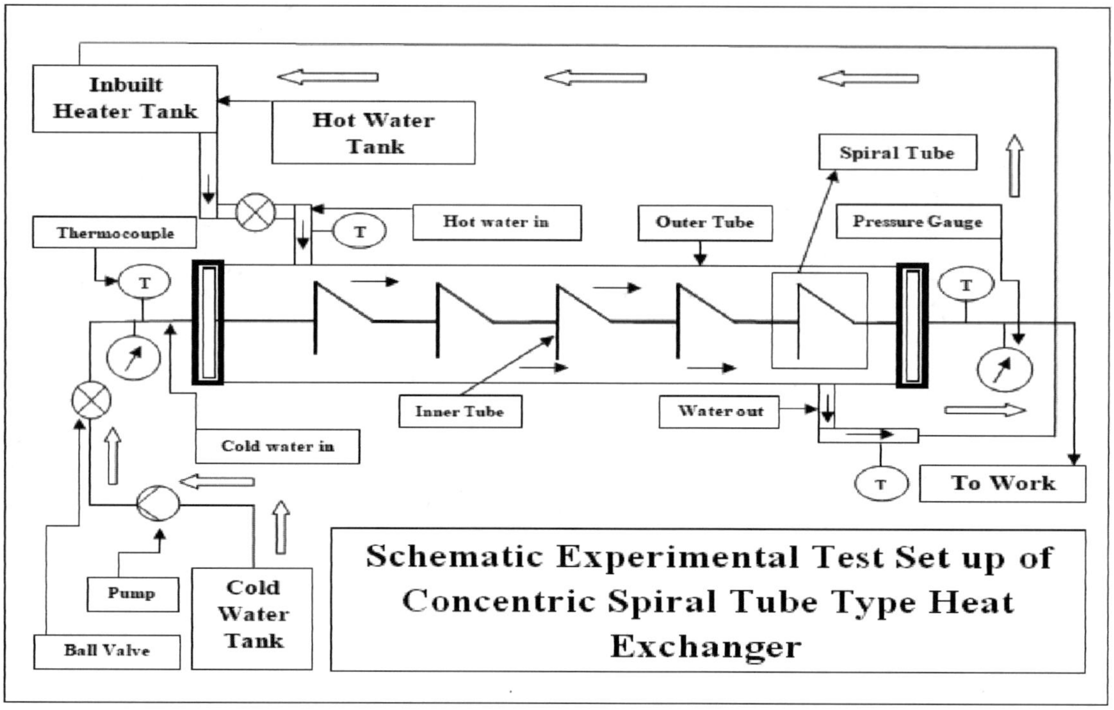

Figure Number 3.1: Schematic diagram of the experimental setup

(a) (b) (c) (d)

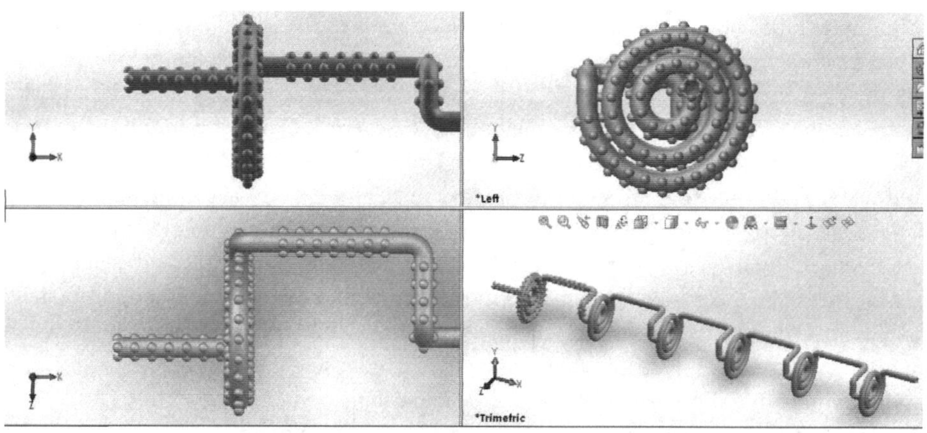

(e)

Figure Number 3.2: (a) Side view of conical shape insert located over spiral tube. (b) Isometric view of conical shape insert located over spiral tube. (c) Side view of plain profile of spiral tube. (d) Front view of plain profile of spiral tube. (e) Side, Front, Isometric view of hemispherical shape insert located over spiral tube.

Table Number 3.1: Geometrical Dimensions of cascaded spiral tube heat exchanger

Parameters	Dimensions
Outer diameter of the tube, mm	20
Inner diameter of the tube, mm	16
Inner most diameter of the spiral coil tube, mm	10.6
Outer most diameter of the spiral coil tube, mm	12.6
Number of coil turns	3
Number of spiral coils	6
Distance between spiral coils, mm	11.4
Outer diameter of shell,mm	150
Inner diameter of shell,mm	144.12
Length of the shell,mm	850
Diameter of inlet and outlet hole of hot water, mm	15

Table Number 3.2: Experimental conditions of cascaded spiral tube heat exchanger

Parameters	Hot Water	Cold Water
Inlet water temperature, (^{o}C)	60	31
Mass flow rate, (kg/s)	.067	.012-.049
Specific heat c_p, (J/kg. ^{o}C)	4185	4178
Prandtl number (Pr)	2.99	5.3
Thermal conductivity K, (W/m^2. ^{o}C)	653	619
Viscosity μ, (Ns/m^2)	.000474	.000780
Density ρ, (kg/m^3)	983	995

4. Results and Discussions:

It has been observed from the figure number 4.1 that the Reynolds number range from 4236 to 18540 as shown in figure, friction factor shows a decrement for different profiles and inserts and its values varies from 0.00588 to 0.00513 for plane profile of spiral shape copper tube having no inserts, 0.00495 to 0.00427 for conical shape inserts located over spiral type inner copper tube profile and 0.00376 to 0.00305 for spiral type copper tube profile having hemispherical shape inserts, where as the simulated value of friction factor for exchanger also shows an decrement for different profiles and inserts and its values varies from 0.00940 to 0.00404 for plane profile of spiral shape copper tube having no inserts, 0.00491 to 0.00325 for conical shape inserts located over spiral type inner copper tube profile and 0.00320 to 0.00203 for spiral type copper tube profile having hemispherical shape inserts, on the other hand experimental values of Nusselt number shows an enhancement for different shape of inserts and its values varies from 35.05 to 98.52 for plane profile of spiral shape tube, 75.15 to 144.05 for conical shape inserts located over spiral copper tube profile and 45.15 to 123.95 for spiral tube profile having hemispherical shape inserts. On comparing the experimental values of these three profiles hemispherical shape insert shows best and most promising results among other profiles, because of good turbulence in fluid flow due to the reason of insert shape and available surface area, the best and optimum Nusselt number is 144.05 and friction factor is 0.00305 for best suitable mass flow rate is 0.049 Kg/s, where as the simulated value of Nusselt number in exchanger also showed an increment for different profiles and inserts and its values varies from 40.25 to 120.25 for plane profile of spiral shape copper tube having no inserts, 48.15 to 146.85 for conical shape inserts located over spiral type inner copper tube profile and 88.15 to 165.85 for spiral type copper tube profile having hemispherical shape inserts. Thus this shows a close agreement and validates the experimental and simulation values of Nusselt number and friction factor. Further the experimental values were also validated for plain spiral profile for Nusselt number and friction factor from the following governing equations as given below.

Blasius equation for Friction factor [09]

$$f_s = 0.316 \times Re^{-0.25} \tag{22}$$

Dittus–Boelter equation for Nusselt number [10]

$$Nu_s = 0.023.Re^{0.8}.Pr^{0.4} \quad (23)$$

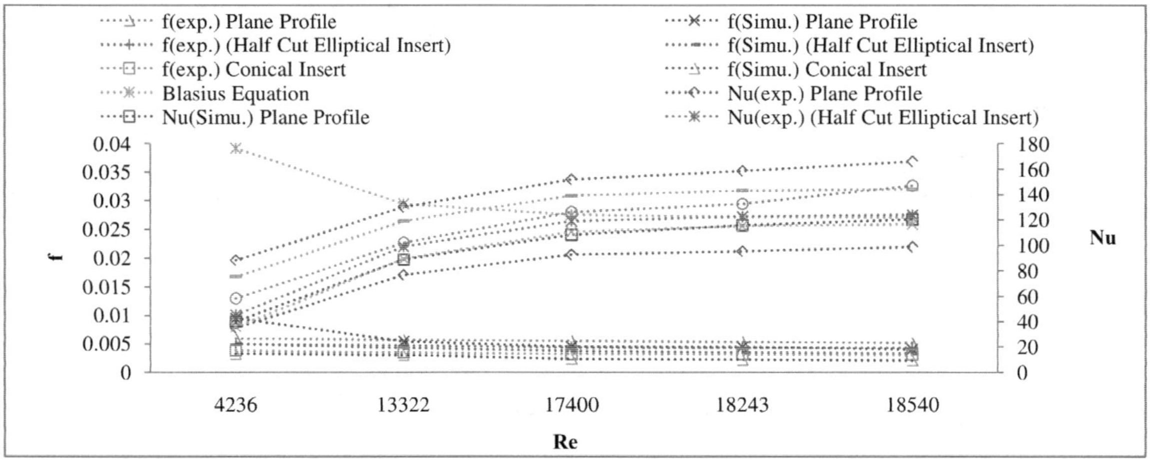

Figure Number 4.1: Variation between Friction Factor and Nusselt Number for varying Reynolds Number.

After comparing both the experimental and simulated values of different profiles like of plane profile of spiral copper tube, spiral coil with conical inserts and spiral coil with hemispherical inserts etc, it has been observed that simulated values were greater than experimental values due to the reason of heat loss occurred from the experimental setup, but it was noted that minimum error percentage occurred between experimental and simulation values which varies from range 2% - 4% for all the properties like Friction factor, Nusselt number, effectiveness and heat transfer rate.

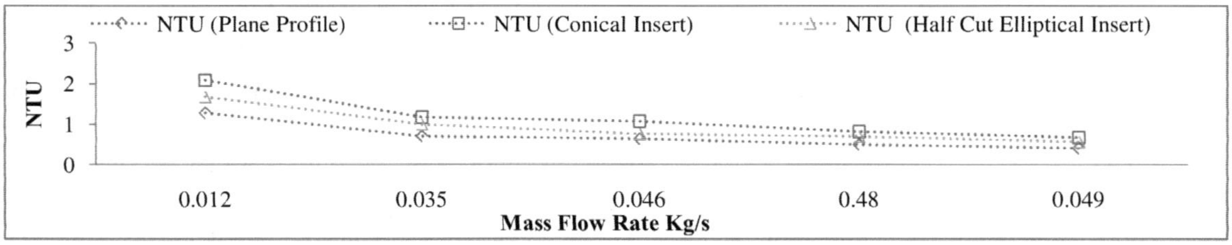

Figure Number 4.2: Variation between Number of Transfer Unit (NTU) and Mass Flow Rate for varying Shapes of Inserts and without Inserts.

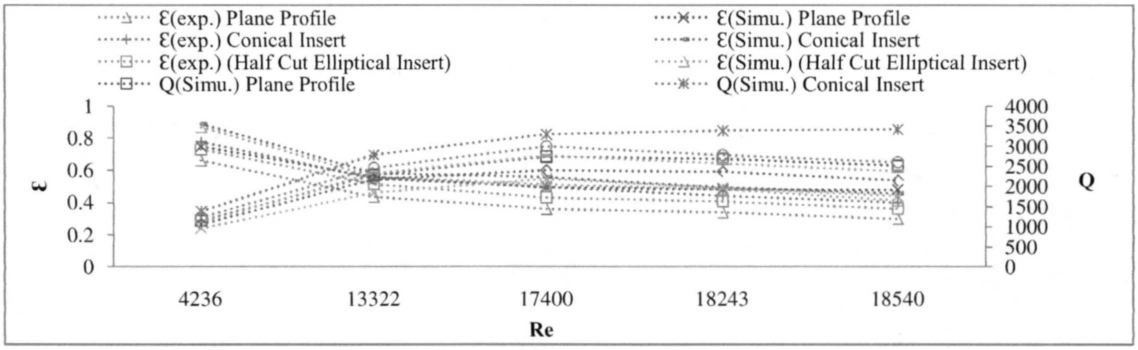

Figure Number 4.3: Variation between Effectiveness and Heat Transfer Rate for varying Reynolds Number.

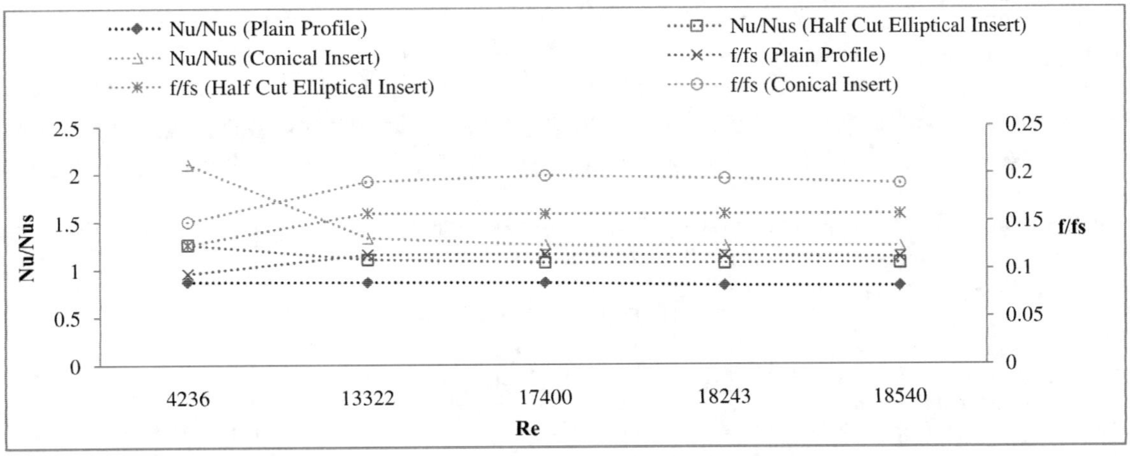

Figure Number 4.4: Variation between Nu/Nu_s and f/f_s with varying Reynolds Number.

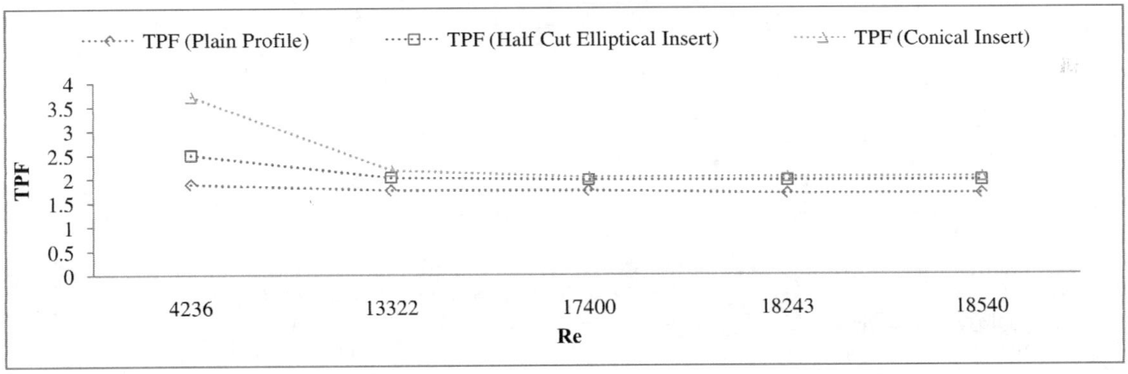

Figure Number 4.5: Variation between Thermal Performance Factor and Reynolds Number for varying Shapes of Inserts and without Inserts.

Thus, this condition of error percentage validated the experimental and simulated values of the experimentation. Now, from the figure number 4.2, it is evidently shown that as the mass flow rate increases from 0.012 to 0.049 Kg/s the value of NTU decreases with range of 1.27 to 0.41 for plane spiral copper tube, 1.67 to 0.56 for spiral tube having conical inserts and 2.08 to 0.68 for spiral copper tube having hemispherical shape insert. On comparing these values, it has been clearly observed that hemispherical inserts located over spiral copper tube showed best and optimum results as compared to other profile due to the reason that hemispherical insert have larger contact area than conical profile and it developed more turbulent region in fluid than other profile due to which number of transfer unit (NTU) occurred more in this shape only. From figure 4.4 and 4.5, it is evidently shown that as the range of Reynolds number increasing from 4236 to18540 the Nu/Nu_s factor f/f_s factor and thermal performance factor is decreasing but conical insert profile show best results as compared to other plain profile and half cut elliptical profile insert.

5. Conclusions:

The simulated and experimental results for the CCSICTHE shows close results and similarity, hence validating each other, in future the results of this cascaded structure can be compared using an analytical approach for similar problem, and also with different designs and shapes of inserts and micro channel over inner tube with diverse fluid and flow patterns in a CTHE. It has been concluded that out of three profiles, i.e. plain, conical and half cut elliptical profile, which located over inner copper tube of exchanger, half cut

IOP Conf. Series: Materials Science and Engineering **802** (2020) 012009 doi:10.1088/1757-899X/802/1/012009

elliptical profile shows best results on fluid flow and thermal peculiarities, for mass flow rate range 0.012 to 0.049 Kg/s and Reynolds number range 4236 to 18540, heat transfer rate varies from 1128.06 to 2374.77, Nusselt number varies from 75.15 to 144.05, Friction factor varies from 0.00376 to 0.00305 and effectiveness varies from 0.77 to 0.40 and Number of transfer unit varies from 2.08 to 0.68 for half cut elliptical profile. The values which was occurred during experimentation, after optimized for evaluating best optimum values for implementation of this research in different areas and from such analysis it is concluded that hemispherical shape insert shows best and most promising as compared to other profiles, The optimum values for mass flow rate is 0.049 Kg/s, Nusselt number is 144.05, friction factor is 0.00305, heat transfer rate is 2374.77 and effectiveness is 0.77.

6. References

[1] P. Naphon, S. Wongwises, 2003, Investigation of the performance of a spiral-coil finned tube heat exchanger under humidifying conditions, J. Eng. Phys. Thermophys. 76, 83–92.

[2] S. Garimella, D.E. Richards, R.N. Christensen, 1998, Experimental Investigation of Heat Transfer in Coiled Annular Ducts, J. Heat Transfer, 110, 329–336.

[3] D.G. Prabhanjan, G.S.V. Raghavan, T.J. Rennie, 2002, Comparison of heat transfer rates between a straight tube heat exchanger and a helically coiled heat exchanger, Int. Comm. Heat Mass Transfer 29, 185–191.

[4] J.C. Ho, N.E. Wijeysundera, "S. Rajasekar, T.T. Chandratilleke, 1995, Performance of a compact spiral coil heat exchanger", Heat Recovery Syst. & CHP 15, 457–468.

[5] J.C. Ho, N.E. Wijeysundera, 1996, Study of a compact spiral-coil cooling and dehumidifying heat exchanger unit, Appl. Therm. Eng. 16, 777–790.

[6] J.C. Ho, N.E. Wijeysundera, 1999, An unmixed-air flow model of a spiral cooling dehumidifying heat transfer, Appl. Therm. Eng. 19, 865–883.

[7] M. Bhola, V. Kumar, Dr. S. Singh, 2015 "Analysis of Heat Transfer Enhancement in Circular Double Tube Heat Exchanger Using Insert in ANSYS FLUENT 14.5",International Journal of Innovative Science, Engineering & Technology, 2, 2348-7968.

[8] M. Bhola, V. Kumar, Dr. S. Singh, 2015, Analysis of Heat Transfer Enhancement in Circular Double Tube Heat Exchanger Using Insert in ANSYS FLUENT, International Journal of Engineering Research and Technology, 04, 2278-0181.

[9] H. Blasius, Grenzschichten in Flussigkeitenmitkleiner Reibung, 1908. (German) Z. Math. Phys. 56, 1–37.

[10] E. Manay, B. Sahin, E.F. Akyürek, Ö. Çomaklı, 2012, Use of nanofluids in micro channels, Engineer and Machine, 627(53), 28–42.

[11] H. Bas, V. Ozceyhan, 2012, Heat transfer enhancement in a tube with twisted tape inserts placed separately from the tube wall, Exp. Therm. Fluid Sci. 4, 51–58.

[12] M.M.K. Bhuiya, A.S.M. Sayem, M. Islam, M.S.U. Chowdhury, M. Shahabuddin, 2014, Performance assessment in a heat exchanger tube fitted with double counter twisted tape inserts, Int. Commun. Heat Mass Transf. 50, 25–33.

[13] M.M.K. Bhuiya, M.S.U. Chowdhury, M. Shahabuddin, M. Saha, L.A. Memon, 2013, Thermal characteristics in a heat exchanger tube fitted with triple twisted tape inserts, Int. Commun. Heat Mass Transf. 48, 124–132.

[14] S. Eiamsa-ard, V. Kongkaitpaiboon, K. Nanan, 2013, Thermohydraulics of turbulent flow through heat exchanger tubes fitted with circular-rings and twisted tapes,Chin. J. Chem. Eng. 21 (6), 585–593.

[15] K. Nanan, C. Thianpong, P. Promvonge, S. Eiamsa-ard, 2014, Investigation of heat transfer enhancement by perforated helical twisted-tapes, Int. Commun. Heat Mass Transf. 52, 106–112.

[16] Nishant Kumara, Shriram S. Sonawanea,∗, Shirish H. Sonawane, 2018, Experimental study of thermal conductivity, heat transfer and friction factor of Al2O3 based nanofluids, International Communications in Heat and Mass Transfer 90, 1–10.

[17] G. Kumaresan, S. Venkatachalapathy, Lazarus Godson Asirvatham , Somchai Wongwises, 2014, Comparative study on heat transfer characteristics of sintered and mesh wick heat pipes using CuO Nanofluids, International Communications in Heat and Mass Transfer 57, 208–215.

[18] Shivasheesh Kaushik, Satyendra Singh, 2019, Analysis on Heat Transmission and Fluid Flow Attributes in Solar Air Accumulator Passage with Diverse Faux Jaggedness Silhouettes on Absorber Panel, International Journal of Engineering and Advanced Technology, 8, 32-41.

Influence of Friction Stir Welding Parameters on Mechanical Properties of dissimilar AA 7475 to AISI 304

Pankul Goel[1*], Ankit Kumar Saxena[1], Mohd Atif Wahid[2], Sachin Rathore[3], Nidhi Sharma[4], Radha Kant Mishra[2]

[1]IMS Engineering College, Department of Mechanical Engineering, Ghaziabad-201009, India.
[2]Delhi Technical Campus, Department of Mechanical Engineering, Greater Noida-201306, India.
[3]KIET Group of Institutions, Department of Mechanical Engineering, Ghaziabad-201026, India.
[4]Greater Noida Institute of Technology, Department of Mechanical Engineering, Greater Noida-201306, India.

Email: pankul.goel@imsec.ac.in

Abstract. Joining of dissimilar materials is difficult and challenging. Trend towards replacement of high density material (HDM) by low density material (LDM) is gaining importance in automotive industries. In the present work, friction stir welding (FSW) of high strength dissimilar materials i.e. aluminium alloy AA7475-T761 and stainless steel AISI 304 was conducted. The mechanical properties and microstructure of butt joint were investigated on using Taguchi L_8 orthogonal array (OA). Different combinations of FSW parameters were used to obtain the maximum value of UTS. It was observed that most dominating parameter affecting UTS is tool rotational speed, contributing 42.19%, followed by traverse speed with 38.96% and shoulder diameter with 12.60%. The maximum UTS was observed 75.22% of base material AA 7475 (452MPa) under the optimum combination of parameters; tool rotational speed at 450 rpm, traverse speed 63 mm/min and shoulder diameter 14 mm. The microstructure of the resulted efficient joint shows appreciable variation in grain sizes in different zones and the micro hardness was found maximum in SZ on the retreating side.

Keywords: Aluminum, Friction Stir Welding (FSW), Mechanical Properties, Microstructure, Microhardness, Steel

1. Introduction

Recent days various transportation industries are focusing and adopting the low density material (LDM) by replacing high density material (HDM) without compromising or degrading the safety and service issues [1-2]. The LDMs are suitable as they offer low inertia and lead to fuel economy. Among LDMs, aluminium alloy (AA) finds extensive application in aerospace and transportation industries. In addition, AA offers numerous features such as good forming ability, anti-corrosive and low absorbing energy, etc. Apart LDMs, stainless steel (SS) one of HDMs, finds substantial applications in exhaust parts and fuel tanks of automotive industries [3]. In order to decline inertia and enhancing of mechanical properties, AA and SS were joined by using a green technology FSW. In 1991, FSW was established as well as patented by "The Welding Institute" (TWI), UK [4]. It is a solid-state welding process due to which it avoids number of quality issues such as distortion, porosity, alteration of microstructure and solidification cracks, etc., inevitable in fusion welding process [5]. FSW process is conducted using a competent fixture where both base material (BM) plates are rigidly clamped with suitable joint configuration such as butt, lap and T joint, etc. A rotating tool (non consumable) with suitable pin profile is plunged into the butt joint of the clamped plates. Rubbing action between tool shoulder and BMs generates heat; followed by plastic deformation due to stirring action of tool pin induces sufficient heat to soften the BMs. Further faying surfaces are joined continuously when tool is traversed along the joint-line.

Nevertheless, limited literature was observed for the feasibility of FSW between AA and SS. Uzun et al. [6] conducted FSW of AA6013-T4 to X5CrNi18-10 stainless steel. They reported fatigue strength 30% lower than of BM AA6013-T6 without showing tensile testing results. Goel et al. [7] investigated materials mixing in friction stir welded (FSWed) joint of AA 7475 and AISI 304. They found that FSW parameters tool diameter 14 mm, tool rotation 560 rpm and traverse speed 50 mm/min renders 71% joint efficiency as compared to AA with 7.31% elongation. Jiang et al. [8] joined AA 6061 and AISI 1018 steel using FSW and obtained sound welded joint. Watanabe et al. [9] investigated joining of AA5083 and SS400 mild steel and observed maximum UTS of FSWed joint approximately to 14% lower than of BM AA5083. Habibina et al. [10] analyzed the effects of FSW parameters on AA 5050 to SS 304 and obtained joints of superior qualities by decreasing tool rotational speed 710 to 500 rpm simultaneously raising the traverse speed 40 to 80 mm/min. In addition, tool offset (up to 1.5mm) results higher UTS by avoiding voids and other defects in the SZ. Chen [11] performed optimization on FSW parameters to joining of AA 6061and SS 400 low-carbon steel. He found good quality of UTS as well as impact values at transverse speed of 0.9 mm/s and rotational speed of 550 rpm. It is concluded from the available literature that sound quality of joint exclusively depends on FSW process parameters (PP).

Despite numerous advantages of FSW of HDMs, very few researches pertaining to high strength alloys are reported in literature. Most of the researches that have been performed are motivated to joining of LDMs i.e. AA5xxx, 6xxx series with AISI 304. 2xxx and 7xxx AAs and negligible research are reported on marine-grade AAs. The vast applications and scarcity of literature motivated us to analyze the mechanism of joining between dissimilar materials. Thus, FSW is performed on AA 7475 with AISI 304 SS with butt joint configuration. The metallurgical characteristics and tensile strength of joints were analyzed and presented. Taguchi's L_8 OA is used to explore the effect of three FSW parameters viz tool rotational speed, traverse speed and shoulder diameter.

2. Experimental Procedure

In this study, AA 7475 and AISI 304 were selected as BMs. Both plates of dimension $180 \times 45 \times 2.5$ mm were used with butt joint configuration to conduct FSW. SS and AA were placed on advancing side (AS) and retreating side (RS) respectively. The UTS of the steel and AA was 505 N/mm^2 and 468 N/mm^2 respectively. A non consumable tool of material tungsten carbide consisting cylindrical pin of diameter 4.5 mm and length 1.7 mm was used to conduct all experiments. Apart those, other parameters such as tool to plate tilt angle of 1.5°, plunge depth of 2.2 mm and tool offset were kept fixed in all experiments. The tool offset was maintained at 1.5 mm towards AA being weaker material in order to avoid defects of tunneling and kissing bond [12]. Subsequently, acetone cleaned plates (dried condition) were placed over backing plate and clamped in the sturdy fixture. Thereafter, a retrofitted robust Vertical Milling Machine (Make: Bharat Fritz Werner, India) was used to conduct all experiments. Using wire electrical discharge machine, coupons of all experiments were cut according to ASTM E8m standard. A tensometer was used to compute UTS of all specimens. Metallographic investigation was performed on specimens using metallurgical microscope. Microhardness was evaluated using hardness testing machine at the joint mid-section by indenting a load of 2 N at 0.5mm indentation distance and delay time of 20 s.

The range of FSW PP used for experimentation is shown in Table 1. Taguchi's L_8 OA is considered for performing the eight experiments (shown in Table 2). Three FSW parameters viz tool rotational speed, traverse speed and shoulder diameters were considered at two levels. Subsequently, the UTS of eight joints were measured and their corresponding signal-to-noise (S/N) ratio were evaluated statistically. Using analysis of mean (ANOM) and analysis of variance (ANOVA) the optimal FSW process parameter arrangement was determined to obtain the maximum UTS.

Table 1. FSW process parameters and their levels.

FSW Process parameters	Symbol	Units	Levels	
			Level 1	Level 2
Tool rotational speed	A	rpm	450	560
Traverse speed	B	mm/min	40	63
Shoulder diameter	C	mm	12	14

Table 2. UTS and S/N ratio values.

Expt. No.	Tool rotational speed (A)	Traverse speed (B)	Shoulder diameter (C)	UTS (MPa)	S/N ratio
1	1	1	1	260.0	48.30
2	1	1	2	259.4	48.28
3	1	2	1	301.0	49.57
4	1	2	2	340.0	50.63
5	2	1	1	180.0	45.11
6	2	1	2	230.0	47.23
7	2	2	1	240.0	47.60
8	2	2	2	276.0	48.82

3. Results and discussion

FSW experimentation under different combination of parameters was successfully conducted. Minitab software was used to analyze the collected UTS data. The FSW parameters optimization to attain high UTS is described in the following sections. Microstructure and microhardness of the sound joint (sample 4) resulted under optimal combinations of FSW PPs are detailed further.

3.1. Analysis of UTS

Effects of three PP were scrutinized on UTS through S/N ratio and ANOM techniques. In order to maximize UTS, "higher-the-better" characteristic is opted to assess the S/N ratio. All the values of UTS with S/N ratio for different levels of FSW PP are available in Table 2. The ANOM values for UTS are given in Table 3.

Table 3. Response Table for S/N ratio (UTS)

FSW Process parameters	Symbol	Units	Levels		Max-Min	Rank
			Level 1	Level 2		
Tool rotational speed	A	rpm	49.19	47.19	2	1
Traverse speed	B	mm/min	47.23	49.16	1.93	2
Shoulder diameter	C	mm	47.65	48.74	1.09	3

Table 4. The ANOVA table for UTS

Source	Sum of Squares	DF	Mean Square	F Value	P value	% Contribution
A	8.03	1	8.03	26.99	0.036	42.19
B	7.42	1	7.42	24.33	0.039	38.96
C	2.40	1	2.40	8.06	0.112	12.60
Residual	1.19	4	0.2977			6.25
Total	19.045	7	18.15			

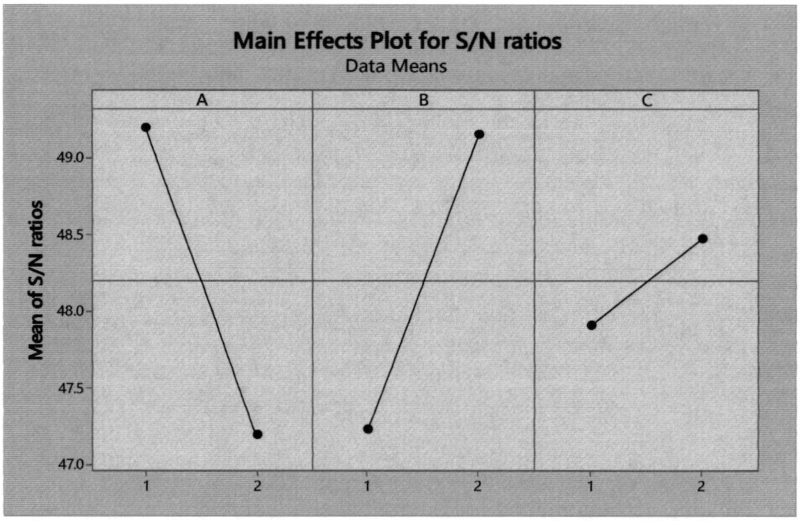

Figure 1. Main effect plots for S/N ratio (UTS).

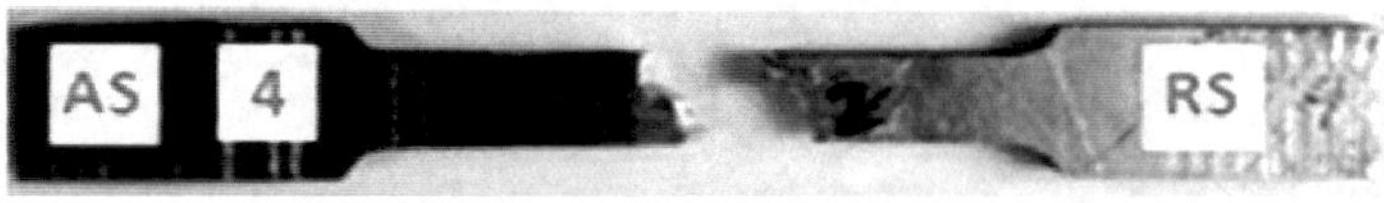

Figure 2. Fractured tensile test coupon.

On the basis of ANOM (Table 3) it can be validated that the UTS is highly susceptible to changes in rotational speed (A). From the main effect plot (Figure 1), the optimum combination of parameters yielding maximum UTS is $A_1B_2C_2$, i.e. tool rotational speed (A) at 450 rpm, traverse speed (B) 63 mm/min and shoulder diameter (C) 14 mm. The maximum UTS was observed 75.22% of base material AA 7475.

ANOVA was used to extricate the relevant conclusion from UTS values as shown in Table 4. Values of prob > P < 0.05, between 0.05 and 0.1 and above 0.1 indicate that parameters are most significant, significant and insignificant, respectively. The tool rotational speed and traverse speed were found as most significant PP (P < .05). Moreover, the percentage contribution of each parameter is evaluated to determine its importance on UTS. Table 4 shows that tool rotational speed (A) contributes maximum towards UTS by 42.19%. Another parameter, traverse speed (B) is also found as good contributor i.e. 38.96% to UTS. However, third parameter tool shoulder diameter (C) contributes marginally towards UTS (only 12.60%).

In order to establish the relation between FSW parameters and UTS, a regression analysis was also performed. This analysis delivers the relationship between process parameters and their response in terms of equation which delivers the response value (UTS). The prediction equation of the UTS in terms of existing FSW parameters was obtained as given in equation (1):

$$UTS_{avg} = 310.8 - 141.1A + 107.9B - 43.3C - 7.8\,AB - 62.8\,AC + 26.2\,BC \qquad (1)$$

The significance of the relationship between UTS and existing parameters is characterized by the value of coefficient of determination i.e. R- squared. Using equation (1), a high value of R- squared is obtained 99.54% which shows that variation in the value of UTS up to 99.54% is explained by the variation of FSW parameters values and rest very small variation (0.46%) in the value of UTS is not clarified by variation of FSW parameters values. Hence this states that a strong relationship exist between UTS and FSW parameters.

Apart this it is evident from Fig.1, UTS decreases when tool rotational speed is increased from level 1 to level 2. Higher tool rotational speed (level 2) might have generated large amount of heat which might have caused formation of IMC and grain growth [13]. However, lower tool rotational speed (level 1) might have produced sufficient heat for the consolidation of soften materials and dynamic recrystallization producing joints with fine grains [14]. The UTS increases as the value of traverse speed raises from level 1 to level 2. At very high traverse speed, tool spends lesser time of stirring due to which low amount of heat is generated. This may result in improper mixing and consolidation of the materials which may decrease strength. However at low traverse speed, material is exposed to considerable stirring action and results in high amount of heat generation along the weld line due to severe plastic deformation. Moreover, the weld zone is exposed to longer duration of high temperature heat causes formation of IMCs, grain growth which ultimately reduces the joint strength. Thus traverse speed at level 2, within the selected range of traverse speed, perhaps induced optimum generation of heat for adequate mixing of BMs in the weld zone and offered better UTS. Shoulder diameter at level 1 might not have generated sufficient heat input and caused inadequate material movement and mixing of dissimilar materials which resulted to decrease of UTS. Increase in shoulder diameter increases the area of contact with BMs. This raises the frictional heat which might be proper to soften SS and consolidates the mixing of BMs and render increment in UTS of FSWed joint [15].

3.2. Microstructure

The microstructure of AS and RS is characterized in three different zones i.e. first SZ, second thermo-mechanically affected zone (TMAZ) and third one heat-affected zone (HAZ). Micrographs of sample 4 (yielding maximum strength at optimized parameters) are shown in Figure 3. Appreciable variation in grain size is visible from the given below Figure 3. From the micrographs it is observed that fine equiaxed grain formation takes place which advocates that grain refinement has been taken substantially in the SZ due to continuous dynamic recrystallization (CDRX). In the TMAZ, the heat and strain remains less than the SZ leading to distorted /partially recrystalized along the welding direction. The highly distorted grains are observed at RS of TMAZ. Since HAZ experiences only thermal cycles, grains are found with large size as well as coarser type. The grains are generally smaller in AS side due to decrease in strain deformation rate. Coarser grain structure leading to lower strength might have been the reason for fracturing of the sample (Fig. 2) from the RS.

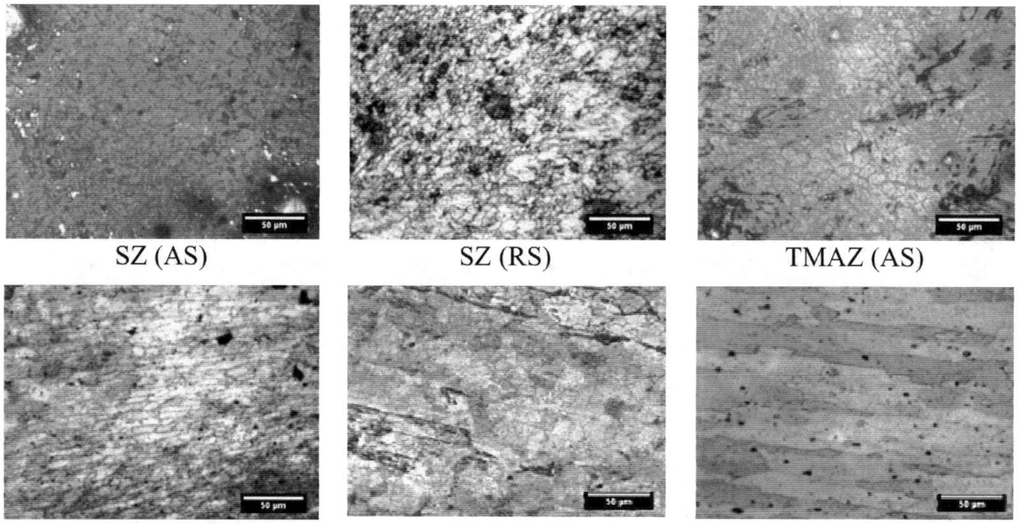

SZ (AS) SZ (RS) TMAZ (AS)

TMAZ (RS) HAZ (AS) HAZ (RS)

Figure 3. Microstructure of different zones of sample 4.

3.3. Microhardness

The average micro-hardness of BM was observed as 164 HV for AA 7475 and 243 HV for AISI 304. The nature of micro-hardness across the FSWed joint is characterized by the joint integrity and microstructure at different zones. Microhardness profile reveals that the maximum hardness occurs at the interface zone, which was assumed to be the presence of brittle IMCs. Microhardness distribution also implies that the considerable grain refinement takes place in the SZ due to CDRX in FSW (Figure 4). Thus, highest micro-hardness is found in SZ of RS, compared to TMAZ/HAZ. Being free from thermo-mechanical action, HAZ is exposed to heat only due to which coarse grains are formed. The resulted coarse grains attribute to low microhardness values.

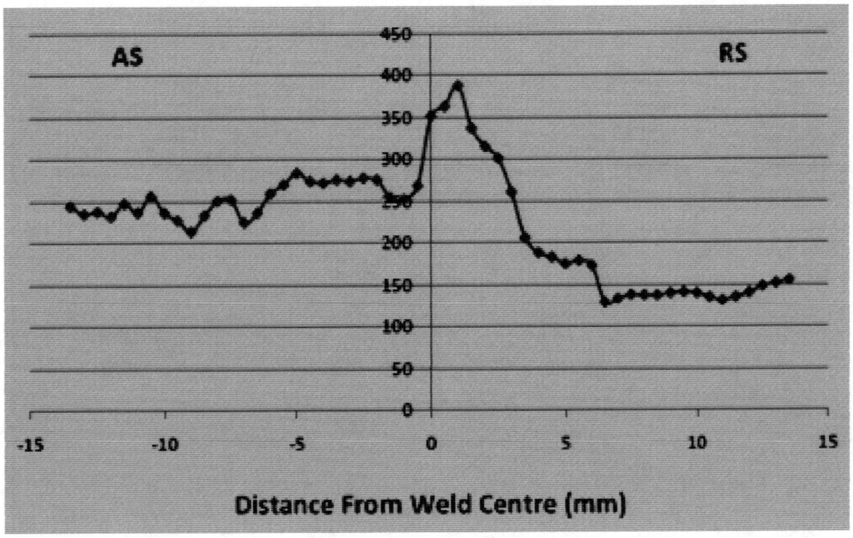

Figure 4. Variation of Micro-hardness of Sample 4.

4. Conclusion

FSW experimentation and optimization of welding parameters were successfully conducted on dissimilar materials AA7475-T761 and AISI 304 using Taguchi L_8 OA.

The results are summarized as follows:

- FSW is a competent process for joining of dissimilar materials AA to SS.
- Increment of traverse speed and decrement of tool rotational speed leads to deliver high UTS..
- UTS is largely influenced by tool rotational speed (contributes 41.29%) successively effected by shoulder diameter (contributes 38.96%).
- The optimum combination of the FSW PP is obtained as $A_1B_2C_2$, i.e., tool rotational speed 450 rpm, traverse speed 63 mm/min and shoulder diameter 14 mm which delivers maximum UTS.
- Optimized parameter combination yields to the maximum value of UTS 340MPa (75% of softer BM AA 7475).
- Significant grain refinement occurs in SZ on both AS and RS. However TMAZ of AS also confirms the considerable grain refinement. This feature attributes to the maximum strength of the joint.
- HAZ was found with non-deformed coarse grains.
- The microhardness of SZ on RS was found greater than the TMAZ, HAZ and BM.

References

[1] Lan, S.H.; Liu, X.; Ni, J. Microstructural evolution during friction stir welding of dissimilar aluminum alloy to advanced high-strength steel. Int. J. Adv. Manuf. Technol. 2016, 82, 2183–2193.
[2] Yazdipour, A.; Heidarzadeh, A. Dissimilar butt friction stir welding of Al 5083-H321 and 316L stainless steel alloys. Int. J. Adv. Manuf. Technol. 2016, 87, 3105–3112.
[3] Goel P., Mohd A.W., Sharma N., Siddiquee A.N., Khan Z.A. (2019) Effects of Welding Parameters in Friction Stir Welding of Stainless Steel and Aluminum. In: Shanker K., Shankar R., Sindhwani R. (eds) Advances in Industrial and Production Engineering. Lecture Notes in Mechanical Engineering. Springer, Singapore.
[4] Thomas WM, Nicholas ED, Needham JC, Murch MG, Temple-Smith P, Dawes CJ (1991) Friction stir

butt welding international patent application no. PCT/GB92/02203 and great Britain patent application no. 9125978.8.
[5] Wahid M.A., Khan, Z.A, Siddiquee AN. Review on underwater friction stir welding: A variant of friction stir welding with great potential of improving joint properties. Transactions of Nonferrous Met.Soc.China. 2018;28(2):193–119.
[6] Uzun H, Donne CD, Argagnotto A, Ghidini T, Gambaro C (2005) Friction stir welding of dissimilar Al 6013- T4 to X5CrNi18-10 stainless steel. Mater Des 26:41–46.
[7] Pankul Goel, Noor Zaman Khan, Zahid A. Khan, Abdulrahman Ahmari, Namrata Gangil, Mustufa Haider Abidi & Arshad Noor Siddiquee (2018): Investigation on material mixing during FSW of AA7475 to AISI304, Materials and Manufacturing Processes, DOI: 10.1080/10426914.2018.1544717.
[8] Jiang WH, Kovacevic R (2004) Feasibility study of friction stir welding of 6061-T6 aluminium alloy with AISI 1018 steel. Proc Inst Mech Eng B J Eng Manuf 218(10):1323–1331. https://doi.org/10.1243/0954405042323612.
[9] Watanabe T, Takayama H, Yanagisawa A (2006) Joining of aluminum alloy to steel by friction stir welding. J Mater Process Technol 178:342–349.
[10] Habibnia M, Shakeri M, Nourouzi S, Givi MKB (2015) Microstructural and mechanical properties of friction stir welded 5050 Al alloy and 304 stainless steel plates. Int J Adv Manuf Technol 76(5–8):819–829.
[11] Chen T (2009) Process parameters study on FSW joint of dissimilar metals for aluminum–steel. J Mater Sci 44:2573–2580.
[12] Khan, N. Z.; Siddiquee, A. N.; Khan, Z. A.; Shihab, S. K., Investigations on Tunneling and Kissing Bond Defects in FSW Joints for Dissimilar Aluminum Alloys. J. Alloys Compd.. 2015, 648, 360–367. DOI: 10.1016/j.jallcom.2015.06.246.
[13] Wahid, M.A., Siddiquee, A., Khan, Z., Asjad, M.: Friction stir welds of Al alloy-Cu: an investigation on effect of plunge depth. Arch. Mech. Eng. 63, 619–634 (2016).
[14] Wang, Q., Zhao, Y., Yan, K., Lu, S.: Corrosion behavior of spray formed 7055 aluminum alloy joint welded by underwater friction stir welding. Mater. Des. 68, 97–103 (2015).
[15] Elangovan K, Balasubramanian V (2008) Influences of tool pin profile and tool shoulder diameter on the formation of friction stir processing zone in AA6061 aluminium alloy Materials & Design, 29(2):362-373.

A SHORT REVIEW ON MACHINING WITH ULTRASONIC MQL METHOD

Rahul Katna[1], Narayan Agrawal[2], M.Suhaib[1]

1. Jamia Millia Islamia, Department of Mechanical Engineering, Faculty of Engineering and Technology, New Delhi
2. Delhi Institute of Tool Engineering, Department of Tool Engineering, Okhla, New Delhi

Email: n _nsit@yahoo.co.in

Abstract:

Cutting fluids are required in huge amount in the modern machining methods. Cutting fluids are needed for providing lubrication as well as cooling the workpiece during machining. Cutting fluids are made of mineral oils and have many drawbacks including many health hazards and environment impact. Various other methods of cutting fluid delivery are being tested in order to reduce the effects of cutting fluids in machining. One such method is minimum quantity lubrication. Many researchers have shown the effectiveness of MQL over the conventional flood method. However the MQL method has drawbacks in terms of heat carrying capacity. In order to increase the effectiveness of the MQL method many advanced methods are being tested for improving the cooling efficiency of the MQL method. One of the methods is ultrasonic method. This method is currently in nascent stages and is being researched upon as a viable alternative to the conventional MQL method. This paper describes this hybrid delivery method in machining.

Keywords: Green; manufacturing; Minimum; quantity; lubrication; Ultrasonic; nanoparticle

1. Introduction

MQL or Minimum Quantity Lubrication is slowly gaining attention from researchers worldwide due to its better lubrication properties and considerably low quantity of cutting fluid[1–5]. MQL has many advantages like it increases tool life and surface finish[6–13]. A typical MQL involves use of high pressure liquid from 2 bar pressure to 5 bar pressure and lubricant in atomized form as shown in Fig 1. The atomized lubricant is in the range of microns and in combination with the high pressure air is able to penetrate the cutting zone effectively[3,14,15]. Machining using MQL method thus provides better lubricity than the conventional flood cooling method. MQL typically uses 50 ml/hr – 200 ml/hr lubricant. An MQL setup consists of a spray nozzle in which the atomized lubricant is mixed either internally or externally depending upon the type of configuration used[16,17]. Researchers have tested MQL method in different machining methods and have found it to be better than the conventional flood delivery method of cutting fluid delivery. MQL is delivered into the cutting zones as shown in Fig 2.

1.1. MQL in machining

Although MQL has proved to be better than the conventional method, researchers have found that combining the MQL method with some other methods like cold air MQL, cryogenic gas, electrically charging the lubricant has performed better than the conventional MQL method itself[18–24]. One of such emerging and promising research is Ultrasonic MQL or UMQL. The ultrasonic MQL method actually combines three methods – first is the MQL itself i.e. atomized lubricant, second is the nanoparticles mixed in the lubricant and third is the ultrasonic mixing mechanism. The result of these triple parts is an efficient lubricity during machining with a constant homogenous delivery of cutting fluid in to the cutting zone. It has been shown that addition of nano particles increase the lubricity of a lubricant. However it is difficult to keep the nanoparticles dispersed in the lubricant[25–27].

The nanoparticles have a tendency to agglomerate and form lumps within the lubricant bulk. Thus, without proper dispersion of the nanoparticles the whole process will get rendered useless. Thus it is important to keep the nanoparticles dispersed in the lubricant for an effective fluid delivery process[28–31]. If the nanoparticles agglomerate together there will be time patches where the lubricant is delivered without nanoparticles and there will be instances where lumps are delivered into the machining zone will eventually be unable to reach the cutting zone thus failing the whole process. Emulsifiers are used for this purpose[32]. An emulsifier is a chemical compound which makes the dispersion of the nanoparticles in the lubricant easy and for a long time. However addition of emulsifiers in the lubricant has shown to affect the properties of the nanoparticles in the lubricant. Another way of tackling this problem is employing a mechanism that can keep the nanoparticles dispersed in the lubricant. Ultrasonic dispersion is one such method in which the nanoparticles[33–37]. This paper considers the latest developments in the ultrasonic method with nanoparticles. Ultrasonic MQL has been used in grinding operation mostly. For efficiently utilizing in other operations it has to be properly setup. Secondly, nanoparticles are costly affair and are lost during machining, so proper care has to be taken for the toxicity effects of nanoparticles on human. Figure 1 shows a typical MQL proceas.

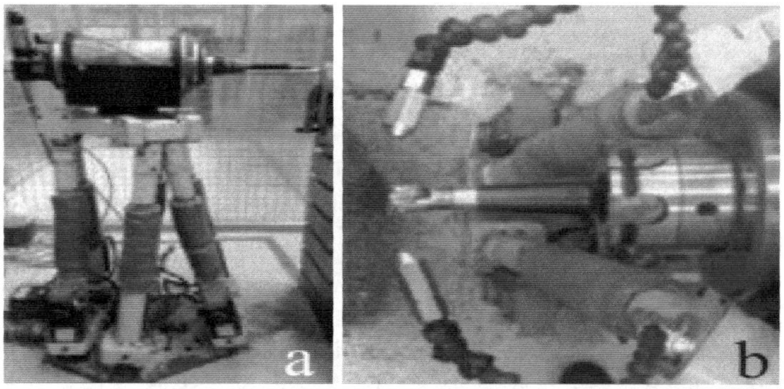

Fig 1: Typical MQL setup[38]

2. Use of hybrid MQL methods in machining with nano particles

In an attempt to increase the thermal efficiency of the MQL method, Rabiei et al [39] used water based nano particles in ultrasonic. They employed six different nanoparticles like oxides of titanium, silicon, aluminium, copper, nickel and multi walled carbon nanotube in grinding. From the experimental results it was observed that there was a reduction of 20% and 24.6% in the grinding force when compared to grinding done without any lubricant or coolant. Also, a superior

IOP Conf. Series: Materials Science and Engineering **802** (2020) 012011 doi:10.1088/1757-899X/802/1/012011

surface finish was achieved without any visible surface defect and damage. Also, no plastic deformation or side flow defect was observed which indicate good lubrication during machining. Interestingly the type of chip obtained with MQL-Ultrasonic grinding was similar to that obtained with the flood method of lubrication which points to the fact that both these methods are similar in terms of machining quality.

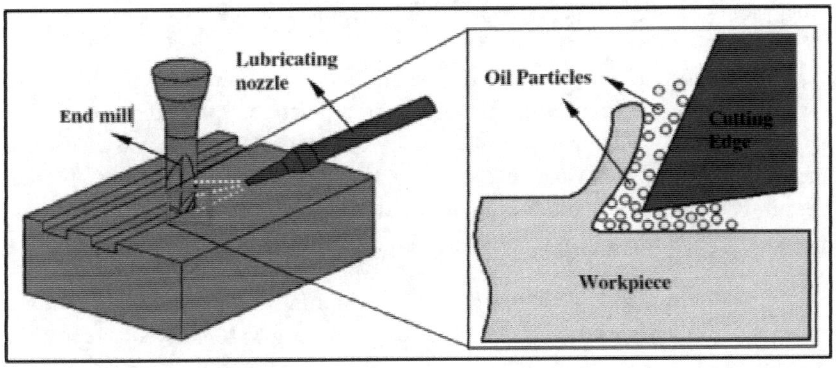

Fig 2: MQL delivery into cutting zone[40]

Huang et al [41] tried to improvise the conventional machining process by using nanoparticles in the cutting fluid and also amalgamating MQL with ultrasonic dispersion. Nanoparticles being solid powders have e tendency to agglomerate in water and cause lumping problems which decreases their performance. An effective way is to use an emulsifier to keep the nanoparticles suspended in the solution. However use of emulsifier increases the cost and can also affect the performance of the nanoparticles. Hence ultrasonic dispersion method was utilized to keep the nanoparticles suspended for a homogenous solution of cutting fluid. In machining mold steel with this method, the experimental results showed that agglomeration of the nanoparticles was reduced to a very large extent. Also, low grinding forces were achieved with this method in comparison to MQL method alone. This hybrid method also showed lower temperature rise and excellent surface finish than the conventional MQL method without nanoparticles and conventional MQL method with nanoparticles.

3. Use of vibration assisted MQL in machining

Ni et al[42] used ultrasonic vibration assisted with the conventional MQL machining method in machining titanium alloy. Introduction of the lubricant during the ultrasonic friction between mating surfaces of the tool and workpiece caused increased lubrication. From the experimental results it was found that the combined method of ultrasonic vibration and MQL method caused reduction in tool wear to a very large extent. On further analyzing the surface of the tool it was found that the major cause of tool failure was fracture of the surface upon impact which was visible on the surface. From the results it can be concluded that the combined method of ultrasonic with MQL is a better way of machining titanium alloy and can also enhance the tool life considerably.

Helmy et al [43] used ultrasonic method of machining combined with MQL method in machining of composite laminates at different machining parameters. The parameters such as cutting speed, feed and depth of cut were varied in machining the composite with a diamond tool and the performance was measure in terms of cutting forces and surface roughness. Experimental results however indicated that flood method of machining produced better results than the MQL method but the MQL method performed nearly equal to the flood method in machining of the composite laminate. This indicates that ultrasonic MQL method is comparable to the flood method and can be improved by changing the machining parameters and optimizing the input parameters.

In another study using oil based nanoparticles Molaie et al used molybdenum disulfide nanoparticles in oil assisted with ultrasonic method in grinding operation[44]. The experimental results clearly indicate that the combined method of ultrasonic grinding clubbed with the conventional MQL method increased the effectiveness of the conventional MQL method and resulted in lower grinding forces. It was also seen from the experimental results that the combined method resulted in lower surface roughness in comparison to the conventionally MQL method which clearly indicates the better lubricity achieved with this hybrid method of fluid delivery.

Madarkar et al [45]used ultrasonic assisted MQL method in machining titanium alloy. An indigenous horn was fabricated in order to produce ultrasonic frequency. The authors used sunflower oil with nanoparticles in 1, 5 and 10% concentration and ultrasonic method of fluid delivery. The combined method of fluid delivery performed better than the conventional MQL method as it yielded lower grinding forces than the conventional MQL method. However, the conventional method yielded lower surface roughness than the ultrasonic assisted grinding process which was due to enhancement in the capacity to retain the sharpness over long period of time than that observed in the conventional MQL method of fluid delivery. Experimental results, however, prove the enhancement in the grindability of titanium alloy with the hybrid method involving ultrasonic and nanoparticles.

Rasidi et al[46] employed a piezoelectric transducer in micro machining with MQL method. They compared dry method with MQL method combined with the ultrasonic method in each case and at two different flow rate of the MQL fluid. The experimental results obtained show a slight improvement in the surface finish with the MQL assisted with ultrasonic method. However, the hybrid method improved the tool wear significantly. Similarly, Li et al[47] reported significant improvement in the tool life of the cutting tool with the ultrasonic assisted machining with MQL.

Alemayehu et al [48] evaluated the machining performance in turning Inconel 718 with a new hybrid method. The authors employed ultrasonic vibration with MQL method in machining the alloy. The authors reported that with the unique method of hybrid delivery lower cutting forces were achieved. Thus this process saves not only input energy and also produces better quality machined surface.

Isobe et al[49] used carbide drill for drilling with carbide drill vibrating ultrasonically. The authors drilled 302 holes by ultrasonic drilling combined with MQL process. Experimental results showed that the micro drilling method combined with MQL method was able to reduce the deflections occurring in the drill bit. Not only this, the authors achieved higher tool life with tool wear reducing by almost half.

4. Conclusion:

AS seen ultrasonic MQL is a relatively newer field and very little literature is published. There is a huge research scope in MQL machining combined with MQL method in machining. It was seen that most of the studies are done in grinding operation only. However MQL method combined with the ultrasonic machining can be tested in milling and turning operations also. It is recommended to test the effectiveness of MQL in ultrasonic machining on superalloys also.

References:

(1) Benedicto, E.; Carou, D.; Rubio, E. M. Technical, Economic and Environmental Review of the Lubrication/Cooling Systems Used in Machining Processes. *Procedia Eng.* **2017**, *184*, 99–116.
(2) Micro/Nanofluids in Sustainable Machining | IntechOpen https://www.intechopen.com/books/microfluidics-and-nanofluidics/micro-nanofluids-in-sustainable-machining (accessed Nov 23, 2019).
(3) Dixit, U. S.; Sarma, D. K.; Davim, J. P. Machining with Minimal Cutting Fluid. In *Environmentally Friendly Machining*; Dixit, U. S., Sarma, D. K., Davim, J. P., Eds.; SpringerBriefs in Applied Sciences and Technology; Springer US: Boston, MA, 2012; pp 9–17. https://doi.org/10.1007/978-1-4614-2308-9_2.
(4) Paul, S.; Sarkar, S. *Integration of Cryogenic Machining Technologies in Advance Manufacturing Systems*; 2019.
(5) Mannekote, J. K.; Kailas, S. V.; Venkatesh, K.; Kathyayini, N. Environmentally Friendly Functional Fluids from Renewable and Sustainable Sources-A Review. *Renew. Sustain. Energy Rev.* **2018**, *81*, 1787–1801. https://doi.org/10.1016/j.rser.2017.05.274.
(6) Dhar, N. R.; Kamruzzaman, M.; Ahmed, M. Effect of Minimum Quantity Lubrication (MQL) on Tool Wear and Surface Roughness in Turning AISI-4340 Steel. *J. Mater. Process. Technol.* **2006**, *172* (2), 299–304. https://doi.org/10.1016/j.jmatprotec.2005.09.022.
(7) Khan, M. M. A.; Mithu, M. A. H.; Dhar, N. R. Effects of Minimum Quantity Lubrication on Turning AISI 9310 Alloy Steel Using Vegetable Oil-Based Cutting Fluid. *J. Mater. Process. Technol.* **2009**, *209* (15), 5573–5583. https://doi.org/10.1016/j.jmatprotec.2009.05.014.
(8) Dhar, N.; Islam, S.; Kamruzzaman, M. Effect of Minimum Quantity Lubrication (MQL) on Tool Wear, Surface Roughness and Dimensional Deviation in Turning AISI-4340 Steel. *Gazi Univ. J. Sci.* **2010**, *20* (2), 23–32.
(9) Kamata, Y.; Obikawa, T. High Speed MQL Finish-Turning of Inconel 718 with Different Coated Tools. *J. Mater. Process. Technol.* **2007**, *192–193*, 281–286. https://doi.org/10.1016/j.jmatprotec.2007.04.052.
(10) Davim, J. P.; Sreejith, P. S.; Silva, J. Turning of Brasses Using Minimum Quantity of Lubricant (MQL) and Flooded Lubricant Conditions. *Mater. Manuf. Process.* **2007**, *22* (1), 45–50. https://doi.org/10.1080/10426910601015881.
(11) Attanasio, A.; Gelfi, M.; Giardini, C.; Remino, C. Minimal Quantity Lubrication in Turning: Effect on Tool Wear. *Wear* **2006**, *260* (3), 333–338. https://doi.org/10.1016/j.wear.2005.04.024.
(12) Hamdan, A.; Sarhan, A. A. D.; Hamdi, M. An Optimization Method of the Machining Parameters in High-Speed Machining of Stainless Steel Using Coated Carbide Tool for Best Surface Finish. *Int. J. Adv. Manuf. Technol.* **2012**, *58* (1), 81–91. https://doi.org/10.1007/s00170-011-3392-5.
(13) Tasdelen, B.; Wikblom, T.; Ekered, S. Studies on Minimum Quantity Lubrication (MQL) and Air Cooling at Drilling. *J. Mater. Process. Technol.* **2008**, *200* (1), 339–346. https://doi.org/10.1016/j.jmatprotec.2007.09.064.
(14) Ezugwu, E. O. Key Improvements in the Machining of Difficult-to-Cut Aerospace Superalloys. *Int. J. Mach. Tools Manuf.* **2005**, *45* (12), 1353–1367. https://doi.org/10.1016/j.ijmachtools.2005.02.003.

(15) Sadeghi, M. H.; Haddad, M. J.; Tawakoli, T.; Emami, M. Minimal Quantity Lubrication-MQL in Grinding of Ti–6Al–4V Titanium Alloy. *Int. J. Adv. Manuf. Technol.* **2009**, *44* (5), 487–500. https://doi.org/10.1007/s00170-008-1857-y.
(16) Zaman, P. B.; Dhar, N. R. Design and Evaluation of an Embedded Double Jet Nozzle for MQL Delivery Intending Machinability Improvement in Turning Operation. *J. Manuf. Process.* **2019**, *44*, 179–196. https://doi.org/10.1016/j.jmapro.2019.05.047.
(17) Rahim, E. A.; Dorairaju, H. Evaluation of Mist Flow Characteristic and Performance in Minimum Quantity Lubrication (MQL) Machining. *Measurement* **2018**, *123*, 213–225. https://doi.org/10.1016/j.measurement.2018.03.015.
(18) Sanchez, J. A.; Pombo, I.; Alberdi, R.; Izquierdo, B.; Ortega, N.; Plaza, S.; Martinez-Toledano, J. Machining Evaluation of a Hybrid MQL-CO2 Grinding Technology. *J. Clean. Prod.* **2010**, *18* (18), 1840–1849. https://doi.org/10.1016/j.jclepro.2010.07.002.
(19) Pereira, O.; Català, P.; Rodríguez, A.; Ostra, T.; Vivancos, J.; Rivero, A.; López-de-Lacalle, L. N. The Use of Hybrid CO2+MQL in Machining Operations. *Procedia Eng.* **2015**, *132*, 492–499. https://doi.org/10.1016/j.proeng.2015.12.524.
(20) Stachurski, W.; Sawicki, J.; Wójcik, R.; Nadolny, K. Influence of Application of Hybrid MQL-CCA Method of Applying Coolant during Hob Cutter Sharpening on Cutting Blade Surface Condition. *J. Clean. Prod.* **2018**, *171*, 892–910. https://doi.org/10.1016/j.jclepro.2017.10.059.
(21) Wakabayashi, T.; Kuhara, J.; Atsuta, T.; Tsukuda, A.; Sembongi, N.; Shibata, J.; Suda, S. Near-Dry Machining of Titanium Alloy with MQL and Hybrid Mist Supply https://www.scientific.net/KEM.656-657.341 (accessed Nov 23, 2019). https://doi.org/10.4028/www.scientific.net/KEM.656-657.341.
(22) Neugebauer, R.; Drossel, W.; Wertheim, R.; Hochmuth, C.; Dix, M. Resource and Energy Efficiency in Machining Using High-Performance and Hybrid Processes. *Procedia CIRP* **2012**, *1*, 3–16. https://doi.org/10.1016/j.procir.2012.04.002.
(23) Kannan, C.; Ramanujam, R.; Balan, A. S. S. Machinability Studies on Al 7075/BN/Al2O3 Squeeze Cast Hybrid Nanocomposite under Different Machining Environments. *Mater. Manuf. Process.* **2018**, *33* (5), 587–595. https://doi.org/10.1080/10426914.2017.1401718.
(24) Xu, J.; Mkaddem, A.; El Mansori, M. Recent Advances in Drilling Hybrid FRP/Ti Composite: A State-of-the-Art Review. *Compos. Struct.* **2016**, *135*, 316–338. https://doi.org/10.1016/j.compstruct.2015.09.028.
(25) Ilie, F.; Covaliu, C. Tribological Properties of the Lubricant Containing Titanium Dioxide Nanoparticles as an Additive. *Lubricants* **2016**, *4* (2), 12. https://doi.org/10.3390/lubricants4020012.
(26) Kolodziejczyk, L.; Martínez-Martínez, D.; Rojas, T. C.; Fernández, A.; Sánchez-López, J. C. Surface-Modified Pd Nanoparticles as a Superior Additive for Lubrication. *J. Nanoparticle Res.* **2007**, *9* (4), 639–645. https://doi.org/10.1007/s11051-006-9124-3.
(27) Wu, Y. Y.; Tsui, W. C.; Liu, T. C. Experimental Analysis of Tribological Properties of Lubricating Oils with Nanoparticle Additives. *Wear* **2007**, *262* (7), 819–825. https://doi.org/10.1016/j.wear.2006.08.021.
(28) Lee, P.-H.; Nam, J. S.; Li, C.; Lee, S. W. An Experimental Study on Micro-Grinding Process with Nanofluid Minimum Quantity Lubrication (MQL). *Int. J. Precis. Eng. Manuf.* **2012**, *13* (3), 331–338. https://doi.org/10.1007/s12541-012-0042-2.
(29) Kalita, P.; Malshe, A. P.; Arun Kumar, S.; Yoganath, V. G.; Gurumurthy, T. Study of Specific Energy and Friction Coefficient in Minimum Quantity Lubrication Grinding Using Oil-Based Nanolubricants. *J. Manuf. Process.* **2012**, *14* (2), 160–166. https://doi.org/10.1016/j.jmapro.2012.01.001.

(30) Wang, X.; Xu, X.; Choi, S. U. S. Thermal Conductivity of Nanoparticle - Fluid Mixture. *J. Thermophys. Heat Transf.* **1999**, *13* (4), 474–480. https://doi.org/10.2514/2.6486.

(31) Mao, C.; Tang, X.; Zou, H.; Huang, X.; Zhou, Z. Investigation of Grinding Characteristic Using Nanofluid Minimum Quantity Lubrication. *Int. J. Precis. Eng. Manuf.* **2012**, *13* (10), 1745–1752. https://doi.org/10.1007/s12541-012-0229-6.

(32) Amrita, M.; Shariq, S. A.; Manoj; gopal, C. Experimental Investigation on Application of Emulsifier Oil Based Nano Cutting Fluids in Metal Cutting Process. *Procedia Eng.* **2014**, *97*, 115–124. https://doi.org/10.1016/j.proeng.2014.12.231.

(33) Taurozzi, J. S.; Hackley, V. A.; Wiesner, M. R. Ultrasonic Dispersion of Nanoparticles for Environmental, Health and Safety Assessment – Issues and Recommendations. *Nanotoxicology* **2011**, *5* (4), 711–729. https://doi.org/10.3109/17435390.2010.528846.

(34) Chung, S. J.; Leonard, J. P.; Nettleship, I.; Lee, J. K.; Soong, Y.; Martello, D. V.; Chyu, M. K. Characterization of ZnO Nanoparticle Suspension in Water: Effectiveness of Ultrasonic Dispersion. *Powder Technol.* **2009**, *194* (1), 75–80. https://doi.org/10.1016/j.powtec.2009.03.025.

(35) Sato, K.; Li, J.-G.; Kamiya, H.; Ishigaki, T. Ultrasonic Dispersion of TiO2 Nanoparticles in Aqueous Suspension. *J. Am. Ceram. Soc.* **2008**, *91* (8), 2481–2487. https://doi.org/10.1111/j.1551-2916.2008.02493.x.

(36) Sauter, C.; Emin, M. A.; Schuchmann, H. P.; Tavman, S. Influence of Hydrostatic Pressure and Sound Amplitude on the Ultrasound Induced Dispersion and De-Agglomeration of Nanoparticles. *Ultrason. Sonochem.* **2008**, *15* (4), 517–523. https://doi.org/10.1016/j.ultsonch.2007.08.010.

(37) Xiaochun, L.; Yang, Y.; Yang, Y.; Weiss, D.; Weiss, D. Theoretical and Experimental Study on Ultrasonic Dispersion of Nanoparticles for Strengthening Cast Aluminum Alloy A356. *Metall. Sci. Tecnol.* **2008**, *26* (2).

(38) Tunc, L. T.; Gu, Y.; Burke, M. G. Effects of Minimal Quantity Lubrication (MQL) on Surface Integrity in Robotic Milling of Austenitic Stainless Steel. *Procedia CIRP* **2016**, *45*, 215–218. https://doi.org/10.1016/j.procir.2016.02.337.

(39) Rabiei, F.; Rahimi, A. R.; Hadad, M. J.; Saberi, A. Experimental Evaluation of Coolant-Lubricant Properties of Nanofluids in Ultrasonic Assistant MQL Grinding. *Int. J. Adv. Manuf. Technol.* **2017**, *93* (9–12), 3935–3953. https://doi.org/10.1007/s00170-017-0774-3.

(40) Khatri, A.; Jahan, M. P. Investigating Tool Wear Mechanisms in Machining of Ti-6Al-4V in Flood Coolant, Dry and MQL Conditions. *Procedia Manuf.* **2018**, *26*, 434–445. https://doi.org/10.1016/j.promfg.2018.07.051.

(41) Huang, W. T.; Liu, W. S.; Wu, D. H. Investigations into Lubrication in Grinding Processes Using MWCNTs Nanofluids with Ultrasonic-Assisted Dispersion. *J. Clean. Prod.* **2016**, *137*, 1553–1559. https://doi.org/10.1016/j.jclepro.2016.06.038.

(42) Ni, C.; Zhu, L.; Yang, Z. Comparative Investigation of Tool Wear Mechanism and Corresponding Machined Surface Characterization in Feed-Direction Ultrasonic Vibration Assisted Milling of Ti–6Al–4V from Dynamic View. *Wear* **2019**, *436–437*, 203006. https://doi.org/10.1016/j.wear.2019.203006.

(43) Helmy, M. O.; El-Hofy, M. H.; El-Hofy, H. Effect of Cutting Fluid Delivery Method on Ultrasonic Assisted Edge Trimming of Multidirectional CFRP Composites at Different Machining Conditions. *Procedia CIRP* **2018**, *68*, 450–455. https://doi.org/10.1016/j.procir.2017.12.077.

(44) Molaie, M. M.; Akbari, J.; Movahhedy, M. R. Ultrasonic Assisted Grinding Process with Minimum Quantity Lubrication Using Oil-Based Nanofluids. *J. Clean. Prod.* **2016**, *129*, 212–222. https://doi.org/10.1016/j.jclepro.2016.04.080.

(45) Madarkar, R.; Agarwal, S.; Attar, P.; Ghosh, S.; Rao, P. V. Application of Ultrasonic Vibration Assisted MQL in Grinding of Ti–6Al–4V. *Mater. Manuf. Process.* **2018**, *33* (13), 1445–1452. https://doi.org/10.1080/10426914.2017.1415451.

(46) Rasidi, I.; Rahim, E. A.; Ibrahim, A. A.; Maskam, N. A.; Ghani, S. C. The Effect on the Application of Coolant and Ultrasonic Vibration Assisted Micro Milling on Machining Performance https://www.scientific.net/AMM.660.65 (accessed Nov 23, 2019). https://doi.org/10.4028/www.scientific.net/AMM.660.65.

(47) Li, K.-M.; Wang, S.-L. Effect of Tool Wear in Ultrasonic Vibration-Assisted Micro-Milling. *Proc. Inst. Mech. Eng. Part B J. Eng. Manuf.* **2014**, *228* (6), 847–855. https://doi.org/10.1177/0954405413510514.

(48) Alemayehu, H.; Ghosh, S.; Rao, P. V. Evaluation of Cutting Force and Surface Roughness of Inconel 718 Using a Hybrid Ultrasonic Vibration-Assisted Turning and Minimum Quantity Lubrication (MQL). In *Advances in Unconventional Machining and Composites*; Shunmugam, M. S., Kanthababu, M., Eds.; Lecture Notes on Multidisciplinary Industrial Engineering; Springer: Singapore, 2020; pp 325–333. https://doi.org/10.1007/978-981-32-9471-4_26.

(49) Isobe, H.; Hara, K. Improvement of Drill Life for Nickel Super Alloy by Ultrasonic Vibration Machining with Minimum Quantity Lubrication https://www.scientific.net/KEM.625.581 (accessed Nov 23, 2019). https://doi.org/10.4028/www.scientific.net/KEM.625.581.

AUTHOR INDEX

Institute of Physics
Dirac House, Temple Back
Bristol BS1 6BE UK

ISSN: 1757-8981
ISBN 978-1-7138-2673-6
9 781713 826736

CURRAN ASSOCIATES INC.
proceedings.com